D0303058

PRC LIBRARY
00095231

PETERBOROUGH
REGIONAL COLLEGE
LIBRARY
ACC.
No. 00095231
CLASS No.
005.
3
CHECKED

Embedded Software Development for Safety-Critical Systems

Embedded Software Development for Safety-Critical Systems

Chris Hobbs

CRC Press is an imprint of the
Taylor & Francis Group, an **informa** business
AN AUERBACH BOOK

CRC Press
Taylor & Francis Group
6000 Broken Sound Parkway NW, Suite 300
Boca Raton, FL 33487-2742

© 2016 by Taylor & Francis Group, LLC
CRC Press is an imprint of Taylor & Francis Group, an Informa business

No claim to original U.S. Government works

Printed on acid-free paper
Version Date: 20150720

International Standard Book Number-13: 978-1-4987-2670-2 (Hardback)

This book contains information obtained from authentic and highly regarded sources. Reasonable efforts have been made to publish reliable data and information, but the author and publisher cannot assume responsibility for the validity of all materials or the consequences of their use. The authors and publishers have attempted to trace the copyright holders of all material reproduced in this publication and apologize to copyright holders if permission to publish in this form has not been obtained. If any copyright material has not been acknowledged please write and let us know so we may rectify in any future reprint.

Except as permitted under U.S. Copyright Law, no part of this book may be reprinted, reproduced, transmitted, or utilized in any form by any electronic, mechanical, or other means, now known or hereafter invented, including photocopying, microfilming, and recording, or in any information storage or retrieval system, without written permission from the publishers.

For permission to photocopy or use material electronically from this work, please access www.copyright.com (http://www.copyright.com/) or contact the Copyright Clearance Center, Inc. (CCC), 222 Rosewood Drive, Danvers, MA 01923, 978-750-8400. CCC is a not-for-profit organization that provides licenses and registration for a variety of users. For organizations that have been granted a photocopy license by the CCC, a separate system of payment has been arranged.

Trademark Notice: Product or corporate names may be trademarks or registered trademarks, and are used only for identification and explanation without intent to infringe.

Visit the Taylor & Francis Web site at
http://www.taylorandfrancis.com

and the CRC Press Web site at
http://www.crcpress.com

Dedication

For
Alexander, Thomas,
and Edward

πόλλ' οἶδ' ἀλώπηξ, ἀλλ' ἐχῖνος ἓν μέγα

Contents

SECTION II: THE PROJECT

SECTION III: DESIGN PATTERNS

Preface

I have written this book for systems designers, implementers, and verifiers who are experienced in general embedded software development, but who are facing the prospect of delivering a software-based system for a safety-critical application. In particular, the book is aimed at people creating a product that must satisfy one or more of the international standards relating to safety-critical applications — IEC 61508, ISO 26262, EN 50128, IEC 62304 or related standards.

Software and Safety

The media seem to delight in describing radiotherapy machines that have given the wrong doses to patients, aviation disasters and near-misses, rockets that have had to be destroyed because of a missed comma in the FORTRAN code,* and many other such software-related failures.

Less often recorded are the times where, hardware having failed, the software *prevented* a disaster. One example of this is the Airbus that landed in the Hudson River, USA, in January 2009 after the hardware of the engines had failed. Without the continued availability of the flight control software, there would almost certainly have been significant loss of life. The hardware/software divide is therefore not completely one-sided.

I hope that this book provides designers, implementers, and verifiers with some ideas that will help to increase the proportion of incidents when software saved the day.

* Variously attributed to the Mariner and Mercury space craft.

References

All of the techniques described in this book may be further explored through hundreds of academic articles. In order to provide you with a way in, I have provided references at the end of each chapter. With a few whimsical exceptions (e.g., Lardner's 1834 paper on diverse programming and Babbage's 1837 paper on the calculating engine), I have included only references that I have personally found useful as a working software developer.

Some of the papers and books I reference changed my ideas about a topic and, in many cases, caused me to start programming to check whether the concepts were actually useful.

I have to admit that I object to paying money to journals to access published papers, and I believe that all the articles to which I refer in the bibliographies at the end of each chapter can be freely (and legally) downloaded from the Internet. In some cases, e.g., Kálmán's original 1960 paper to which Chapter 8 refers, one has the option of paying the original journal $25 to retrieve the article, or going to the website of the associated educational establishment (in this case, the University of North Carolina, USA) and downloading the paper for free. I leave the choice to the reader.

Most of the documents that I reference are various international standards. When I refer to a standard, in particular to a section number within a standard, I mean the following editions:

ISO 26262	First Edition, 2011
ISO 29119	First Edition, 2013 (parts 1 to 3)
ISO 14971	Second Edition, 2007
IEC 61508	Second Edition, 2010
IEC 62304	First Edition, 2006

Tools

I have used the C programming language in examples because this is the language most strongly associated with embedded programming. Some knowledge of C would therefore be useful for the reader, but the examples are short and should be readable by software engineers familiar with other languages.

From time to time in this book, I mention a particular tool. In general I describe open-source tools, not because these are always superior to commercial tools, but because it would seem invidious to mention one commercial vendor rather than another unless I had carried out a careful comparison of the available products.

Tools are also very personal. We all know that wars can arise from discussions of whether `vi` or `emacs` is the better editor. I refuse to take the bait in these discussions, comfortable in my knowledge that `vi` is far better than `emacs`.

So, my choice of tool may not be your choice of tool. I hope that whenever I mention a tool, I provide enough information for you to seek out commercial vendors.

Acknowledgments

This book has evolved from material used by QNX Software Systems (QSS) for a training module covering tools and techniques for building embedded software systems to be used in safety-critical devices deployed in cars, medical devices, railway systems, industrial automation systems, etc.

I am grateful to QSS management, and in particular my direct managers, Steve Bergwerff and Adam Mallory, for allowing me to develop many of the ideas contained in this book while working for them, and for allowing me to use the training material as the foundation for this book.

The material in this book has emerged from my professional work and, of course, I have not worked alone. There are many people whom I need to thank for providing me with ideas and challenges. These particularly include Rob Paterson, Ernst Munter, John Hudepohl, Will Snipes, John Bell, Martin Lloyd, and Patrick Lee — a stimulating group of people with whom I have worked over the years.

Even with the ideas, creating a book is not a trivial task, and I would like to thank my wife, Alison, who has read every word with a pencil in her hand. I cannot count how many pencils she has worn out getting this book into a shape that she finds acceptable. I thank her. I also thank Chuck Clark and the anonymous reader provided by the publisher for their extremely thorough and detailed proofreading.

Some of the cover pictures were provided by QNX Software Systems and Chuck Clark. Many thanks for the permission to use these photographs.

About the Author

To some extent, I have been shaped by three events.

The first I can date precisely: 24th December 1968, at about 8 o'clock in the evening. I was home from university where I was a first-year undergraduate studying a general pure mathematics course. I had completed the first term of my first year, and before leaving for the vacation, I grabbed a book more or less at random from the library in the mathematics common room. I opened it on Christmas Eve and met Kurt Gödel's incompleteness results for the first time. In those days, these meta-mathematical results were not taught at high school, and the intellectual excitement of Gödel's argument hit me like a juggernaut. I went back after the vacation intent on studying this type of mathematics further and had the incredible opportunity of attending tutorials by Imre Lakatos at the London School of Economics. Even now, when I have a lot more knowledge of the context within which Gödel's paper was published, I still think that these results were the highest peak of intellectual creativity of the 20th century.

The most recent event had a gentler introduction. On a whim in 2002, I decided to learn the 24 songs that comprise Franz Schubert's *Winterreise* song cycle. Thirteen years later, I am still working on them with my wife as accompanist, and, from time to time, feel that I have lifted a lid and been allowed to peep into their enormous emotional and intellectual depth.

The intermediate event is of more direct relevance to this book and is recounted in Anecdote 1 on page 4. This event led me into the study of the development of software for safety-critical systems.

It may seem strange to link Gödel's results, safety-critical software and *Winterreise,* but I feel that each represents a first-rank intellectual challenge.

I met a project manager recently who said that developing software for a safety-critical application is just like developing software for any other application, with the overhead of a lot of paperwork. Perhaps I should have dedicated this book to him.

Chris Hobbs
Ottawa

BACKGROUND I

Chapter 1

Introduction

Dependable, Embedded Software

This is a book about the development of dependable, embedded software.

It is traditional to begin books and articles about embedded software with the statistic of how many more lines of embedded code there are in a modern motor car than in a modern airliner. It is traditional to start books and articles about dependable code with a homily about the penalties of finding bugs late in the development process — the well-known exponential cost curve.

What inhibits me from this approach is that I have read Laurent Bossavit's wonderful book, *The Leprechauns of Software Engineering* (reference [1]), which ruthlessly investigates such "well-known" software engineering preconceptions and exposes their lack of foundation.

In particular, Bossavit points out the circular logic associated with the exponential cost of finding and fixing bugs later in the development process: "Software engineering is a social process, not a naturally occurring one — it therefore has the property that what we believe about software engineering has causal impacts on what is real about software engineering." Because we expect it to be more expensive to fix bugs later in the development process, we have created procedures that make it more expensive.

Bossavit's observation is important and will be invoked several times in this book because I hope to shake your faith in other "leprechauns" associated with embedded software.

Safety Culture

> *A safety culture is a culture that allows the boss to hear bad news.*
>
> Sidney Dekker

Most of this book addresses the technical aspects of building a product that can be certified to a standard, such as IEC 61508 or ISO 26262. There is one additional, critically important aspect of building a product that could affect public safety — the responsibilities carried by the individual designers, implementers and verification engineers. It is easy to read the safety standards mechanically, and treat their requirements as hoops through which the project has to jump, but those standards were written to be read by people working within a safety culture.

Annex B of ISO 26262-2 provides a list of examples indicative of good or poor safety cultures, including "groupthink" (bad), intellectual diversity within the team (good), and a culture of identifying and disclosing potential risks to safety (good). Everyone concerned with the development of a safety-critical device needs to be aware that human life may hang on the quality of the design and implementation.

Anecdote 1 *I first started to think about the safety-critical aspects of a design in the late 1980s when I was managing the development of a piece of telecommunications equipment.*

A programmer, reading the code at his desk, realized that a safety check in our product could be bypassed. The system was designed in such a way that, when a technician was working on the equipment, a high-voltage test was carried out on the external line as a safety measure. If a high voltage was present, the software refused to close the relays that connected the technician's equipment to the line.

The fault found by the programmer allowed the high-voltage check to be omitted under very unusual conditions.

I was under significant pressure from my management to ship the product. It was pointed out that high voltages rarely were *present and, even if they were, it was only under very unusual circumstances that the check would be skipped.*

I now realize that, at that time, I had none of the techniques described in this book for assessing the situation and making a reasoned and justifiable decision. It was this incident that set me off down the road that has led to this book.

The official inquiry into the Deepwater Horizon tragedy (reference [2]) specifically addresses the safety culture within the oil and gas industry: "The immediate causes of the Macondo well blowout can be traced to a series of identifiable mistakes made by BP, Halliburton, and Transocean that reveal such systematic failures in risk management that they place in doubt the safety culture of the entire industry."

The term "safety culture" appears 116 times in the official Nimrod Review (reference [3]) following the investigation into the crash of the Nimrod aircraft XV230 in 2006. In particular, the review includes a whole chapter describing what is required of a safety culture and explicitly states that "The shortcomings in the current airworthiness system in the MOD are manifold and include ... a Safety Culture that has allowed 'business' to eclipse Airworthiness."

In a healthy safety culture, any developer working on a safety-critical product has the right to know how to assess a risk, and has the duty to bring safety considerations forward.

As Les Chambers said in his blog in February 2012* when commenting on the Deepwater Horizon tragedy:

> *We have an ethical duty to come out of our mathematical sandboxes and take more social responsibility for the systems we build — even if this means career threatening conflict with a powerful boss. Knowledge is the traditional currency of engineering, but we must also deal in belief.*

One other question that Chambers addresses in that blog posting is whether it is acceptable to pass a decision "upwards." In the incident described in Anecdote 1, I refused to sign the release documentation and passed the decision to my boss. Would that have absolved me morally or legally from any guilt in the matter, had the equipment been shipped and had an injury resulted? In fact, my boss also refused to sign and shipment was delayed at great expense.

I am reminded of an anecdote told at a conference on safety-critical systems that I attended a few years back. One of the delegates said that he had a friend who was a lawyer. This lawyer quite often defended engineers who had been accused of developing a defective product that had caused serious injury or death. Apparently, the lawyer was usually confident that he could get the engineer proven innocent if the case came to court. But in many cases the case never came to court because

* `http://www.systemsengineeringblog.com/deus_ex_machina/`

the engineer had committed suicide. This anecdote killed the conversation as we all reflected on its implications for each of us personally.

Our Path

I have structured this book as follows:

Background material.

Chapter 2 introduces some of the terminology to be found later in the book. This is important because words such as *fault*, *error*, and *failure*, often used almost interchangeably in everyday life, have much sharper meanings when discussing embedded systems.

A device to be used in a safety-critical application will be developed in accordance with the requirements of an international standard. Reference [4] by John McDermid and Andrew Rae points out that there are several hundred standards related to safety engineering.

From this smörgåsbord, I have chosen a small number for discussion in Chapter 3, in particular IEC 61508, which relates to industrial systems and forms the foundation for many other standards; ISO 26262 which extends and specializes IEC 61508 for systems within cars; and IEC 62304, which covers software in medical devices.

I also mention other standards, for example, IEC 29119, the software testing standard, and EN 50128, the railway standard, to support my arguments here and there in the text.

To make some of the discussion more concrete, in Chapter 4 I introduce two fictitious companies. One of these is producing a device for sale into a safety-critical market, and the other is providing a component for that device. This allows me to illustrate how these two companies might work together within the framework of the standards.

Developing a product for a safety-critical application.

Chapter 5 describes the analyses that are carried out for any such development — a hazard and risk analysis, the safety case analysis, the failure analysis, etc. — and Chapter 6 discusses the problems associated with incorporating external (possibly third-party) components into a safety-critical device.

Techniques recommended in the standards.

Both IEC 61508 and ISO 26262 contain a large number of tables recommending various techniques to be applied during a software development. Many of these are commonly used in any software development (e.g., interface testing), but some are less well-known. Chapters 7 to 19 cover some of these less common techniques.

For convenience, I have divided these into patterns used during the architecture and design of a product (Chapters 7 to 10), the tools that are used to ensure the validity of the design (Chapters 11 to 15), the techniques used during implementation (Chapters 16 to 18), and those used during implementation verification (Chapter 19).

Development tools

One essential, but sometimes overlooked, task during a development is preparing evidence of the correct operation of the tools used; it is irrelevant whether or not the programmers are producing good code if the compiler is compiling it incorrectly. Chapter 20 provides some insights into how such evidence might be collected.

I introduce the goal structuring notation and Bayesian belief networks in Chapter 5, but relegate details about them to Appendices A and B respectively. Finally, there is a little mathematics here and there in this book. Appendix C describes the notations I have used.

Choosing the Techniques to Describe

I have used three criteria to guide me in selecting the tools and techniques to describe:

1. The technique should be explicitly recommended in IEC 61508 or ISO 26262 or both. I think that I have only included one technique not directly recommended in those standards.
2. I should have made direct use of the tool or technique myself.
3. The tool or technique should not be in common use in the majority of companies that I have recently visited.

Development Approach

There are many development methodologies ranging from pure waterfall (to which Bossavit dedicates an entire chapter and appendix in reference [1]) through modified waterfall, design-to-schedule, theory-W, joint application development, rapid application development, timebox development, rapid prototyping, agile development with SCRUM to eXtreme programming.

I do not wish to take sides in the debate about which of these techniques is the most appropriate for the development of an embedded system, and so, throughout this book, I will make use of the composite approach outlined in Figure 1.1.

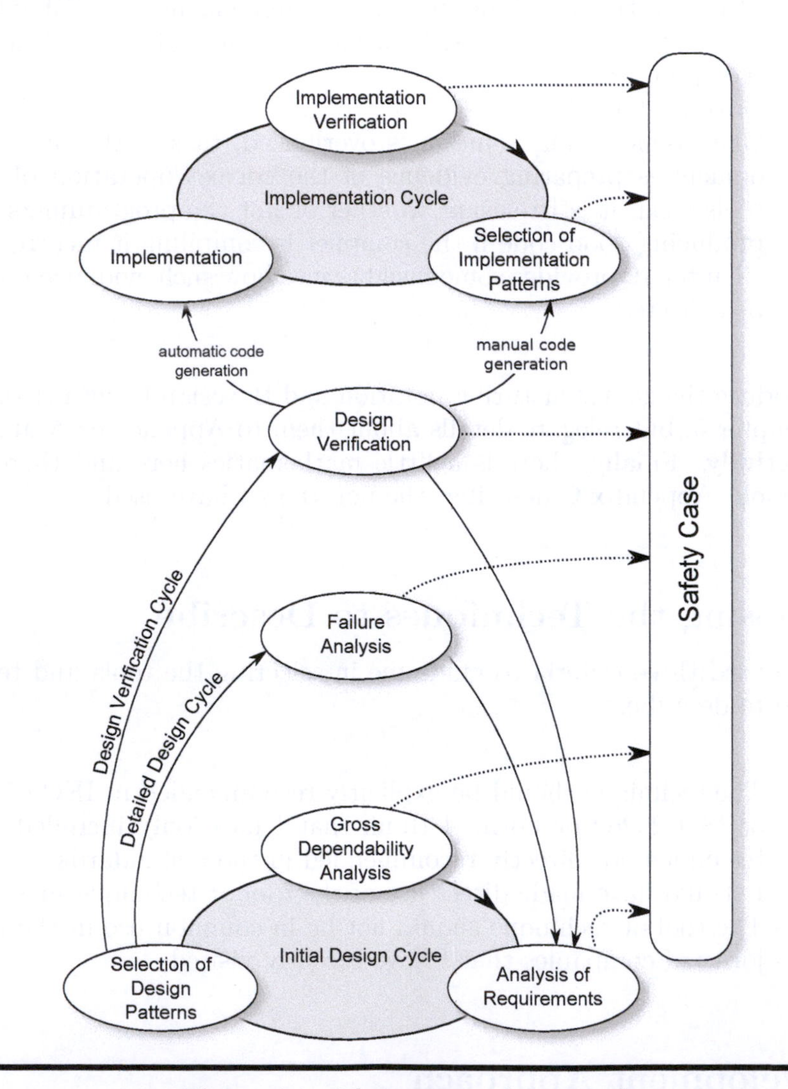

Figure 1.1 The suggested approach.

During a development process, certain tasks have to be completed (although typically not sequentially):

Initial design cycle.
: During the *initial design cycle,* shown at the bottom of Figure 1.1 the requirements for a component or system are analyzed and proposals made for possible designs. These are subject to "quick-and-dirty" analysis (often a Markov Model: see Chapter 11) to see whether they could possibly offer the dependability required to form the basis of the final design.

Detailed design cycle.
: Once one or more candidate designs are selected, a full failure analysis is performed. Whether the failure analysis is carried out by means of a simple Markov model, or by more sophisticated techniques such as fault tree analysis (Chapter 12), the predicted failure rate of the software must be known. Determining the failure rate of software is a thorny issue which I address in Chapter 13.

Design verification cycle.
: When the elements of the design have been selected, the design will need to be verified: Does it handle all possible circumstances, could the device ever lockup and not make progress, etc.?

Implementation cycle.
: Once the design is complete, implementation begins and patterns must be chosen to implement and verify the implementation of the product.

Note that I have carefully avoided saying whether these cycles are followed only once for the development of a particular device (a waterfall process), or are followed for each component, perhaps repeating once every few weeks (an agile process) or even once per day (eXtreme programming).

Given this spectrum of methodologies, I have my personal belief about the end of the spectrum most likely to result in a successful project delivered on-time and on-budget and meeting its safety requirements, but that belief is irrelevant to this book — the techniques I describe are necessary whatever approach is chosen.

However, one point can be noted from the list above: I have placed a lot more emphasis on design and design verification than on implementation and implementation verification. That is deliberate, and in Chapter 15, I try to justify my claim that the skills required for designing and for implementing are sufficiently diverse that it is unlikely that one person can do both.

Another point emphasized by Figure 1.1 is the central role played by the safety case. All of the analyses and decisions made during the project must become elements of the safety case. This essential document is described in Chapter 5, starting on page 61.

Today's Challenges

Although the last ten to fifteen years have seen a tremendous growth in tools and techniques that we can apply to the development of safe, embedded systems, challenges still remain and new ones are emerging.

Security

Safety is now interlinked with security — the behavior of an insecure device is effectively unpredictable once a hacker has broken in. This topic is discussed briefly on page 92, but no general methodology exists for combining security and safety issues. In the past, many embedded systems have been physically secure, being locked in the driver's cab of a train or behind a locked fence on a shop floor. Today almost every device supports external communications through Wi-Fi, Bluetooth, USB drives, or from GPS satellites, and these channels are all security vulnerabilities.

In reference [5], Peter Bernard Ladkin discusses whether the interaction between the company operating a safety-critical system and a malicious attacker can be modeled as a form of game theory, actually a form of meta-game theory, because each protagonist can choose the game to play. He believes that this approach is much more suitable for assessing security threats than the hazard and risk approaches described in the safety standards.

Tools for Balancing Architectural Needs

Chapter 7 identifies the need for an architect and a designer to balance the availability, reliability, performance, usefulness, security, and safety of a system. These characteristics interact, and a design decision in one area will affect all the others. I know of no tools that can help an analyst manipulate these system characteristics and trade one off against the others.

Hardware Error Detection

As described on page 136, processor and memory hardware has become significantly less reliable over the last decade. This has occurred as a result of the increased complexity of chips (most processors come with 20 or more pages of *errata*), and the reduced size of cells that has made them more susceptible to cross-talk, electromagnetic interference and the secondary effects of cosmic rays.

To make matters worse, much of a modern processor, including caches and coprocessors, is hidden from the operating system, and application programs and so errors cannot be detected directly, only through their effect on program operation.

New processors are emerging that offer improved levels of error detection, but these are more expensive and not yet fit-for-purpose.

Deadline Prediction

Many embedded systems in the past have been very simple, and it has been possible to perform real-time calculations to show that, under all circumstances, certain deadlines will be met. As real-time applications* migrate to running alongside other applications on multicore processors with non-deterministic instruction and data caches, these hard guarantees are disappearing and being replaced by guarantees of the form "the probability of deadline X being missed is less than 10^{-6} per hour of operation."

There are few tools beyond discrete-event simulation (see page 190) that allow these sorts of calculations to be performed.

Data

The development of safety-critical systems has only recently begun to take data, particularly configuration data, seriously. Until a few years ago, analysis tended to concentrate on two aspects of a system: hardware and software. It is now becoming increasingly recognized that there is a third element that is equally important: data.

Under the direction of Mike Parsons, the Data Safety Initiative Working Group (DSIWG), which is part of the Safety Critical Systems Club (SCSC), has prepared some very useful guidelines on the use of data in safety-critical systems, but much more work is needed in this area. Annex A of the report presented by the DSIWG to the 2015 Safety Critical Systems Symposium and the associated paper, ref-

---

* For a definition of "real-time," see page 23.

erence [6] by Paul Hampton and Mike Parsons, contains a chilling list of examples where human life has been lost through data errors, rather than hardware or software errors.

References

1. L. Bossavit, *The Leprechauns of Software Engineering: How Folklaw Turns into Fact and What to Do about It.* Leanpub, 2013.
2. B. Graham, W. K. Reilly, F. Beinecke, D. F. Boesch, T. D. Garcia, C. A. Murray, and F. Ulmer, "Deep Water — The Gulf Oil Disaster and the Future of Offshore Drilling: Report to the President," 2011.
3. C. Haddon-Cave, *The Nimrod Review.* UK Government, 2009.
4. J. McDermid and A. Rae, "Goal-Based Safety Standards: Promises and Pitfalls," in *2012 Safety Critical Systems Symposium*, SSS '12, (Bristol, UK), Safety-Critical Systems Club, 2012.
5. P. B. Ladkin, "Risks People Take and Games People Play," in *2015 Safety Critical Systems Symposium*, SSS '15, Safety-Critical Systems Club, 2015.
6. P. Hampton and M. Parsons, "The Data Elephant," in *2015 Safety Critical Systems Symposium*, SSS '15, Safety-Critical Systems Club, 2015.

Chapter 2

Terminology of Safety

To-day we have naming of parts.

Henry Reed

In order to discuss safety-critical software, we need to distinguish between terms that are used almost interchangeably in everyday use. Differentiating, for example, between a "fault," an "error," and a "failure" allows us to tackle them with different techniques; differentiating between "availability" and "reliability" allows us to balance the two in our system. This chapter describes a few of these terms.

General Safety Terminology

Hazard, Risk, Mitigation, and Residual Risk

The iceberg is the *hazard;* the *risk* is that a ship will run into it. One *mitigation* might be to paint the iceberg yellow to make it more visible. Even when painted yellow, there is still the *residual risk* that a collision might happen at night or in fog.

More generally, a hazard is something passive that exists in the environment and which may be the cause of risks. Risks are active and may give rise to dangerous conditions. When a risk has been identified, if it is serious enough, it will need to be mitigated; that is, action will need to be taken to reduce it.

To take an example more pertinent to embedded software, the memory used by the processor may be a hazard. The risks associated with it may include the following:

Risk: The secondary effects of cosmic rays.

These may cause a bit-flip in the configuration memory, changing the device's operating mode unexpectedly and potentially giving rise to a dangerous situation.

One mitigation against this risk might be to use error checking and correcting memory (ECC) so that the errors are detected (and possibly corrected).

Residual risks include the facts that not all errors are detected by ECC memory, that the application may not correctly handle the interrupt indicating the memory error and the fact that ECC may not cover caches.

Risk: The exhaustion of memory.

An application may have a memory leak and, being unable to access additional memory, may fail in an unexpected way, potentially causing a dangerous situation.

One mitigation might be for each application to reserve sufficient memory statically at startup and never release it.

There is still a residual risk that the amount of statically allocated memory is inadequate when a condition occurs that was never considered during design and test.

Usually, there are several risks associated with each hazard, and several mitigations associated with each risk.

Availability, Reliability, and Dependability

When a software subsystem is invoked, it may fail in one of two ways: It may fail to give a timely answer at all, or it may respond, but with the wrong answer. The failure to provide an answer within the time that makes it useful is termed an *availability* problem; the timely presentation of the wrong answer is a *reliability* problem.

Increasing availability generally reduces reliability and *vice versa*. Consider a server that performs some form of calculation. We can increase the reliability of that system by getting two servers to do the calculation and compare their answers. This is a venerable technique:

> *When the formula is very complicated, it may be algebraically arranged for computation in two or more distinct ways, and two or more sets of cards may be made. If the same constants are now employed with each set, and if under these circumstances the results agree, we may be quite secure in the accuracy of them.*
>
> Charles Babbage, 1837 (reference [1])

Unfortunately, although it improves the reliability of some systems, duplication of this nature reduces availability — if either of the two servers is unavailable, then the entire system is unavailable.

Given this antagonism between availability and reliability, it is important during the design process to understand which is to be emphasized in that design and how a balance is to be achieved. This topic is explored further on page 90.

It is relatively easy to think of systems where reliability is essential to safety, even at the expense of availability. It is harder, but possible, to imagine systems where availability is more important. In reference [2], Mats Heimdahl describes a deliberately far-fetched example of a system where any increase in reliability will make the system less safe.

Note that, while the clients of a service can, in general, measure the *availability* of the server they are accessing, it is often impossible for them to measure the server's *reliability* — if the clients knew the correct answer then they presumably wouldn't have needed to invoke the provider.

Much of program testing is really testing reliability, only checking availability in passing.

In this book, I use the term "dependability" to combine the two terms "availability" and "reliability," whichever is the more important for the system under consideration. A common definition of "dependable" is that a system is dependable if it delivers a service that can be justifiably trusted.

Functional Safety

Safety can be provided in different ways. Consider a laser on a piece of telecommunications transmission equipment: such a laser is a hazard. One risk associated with this hazard is that if an operator removed the fiber and looked into the transmitter, eye damage could result.

One way of mitigating this risk might be to install the transmitter facing downward close to the bottom of the rack — the position making it impossible for an operator to align an eye with the transmitter. This would be safe, relying on a passive safety system.

Another way of mitigating the risk would be to have software monitoring the connector and turning the laser off if the fiber were removed. To be safe, this requires an active component, the software, to function correctly. The software must be monitoring the connector all the time and must turn the laser off within a few milliseconds to prevent harm.

The second of these techniques represents *functional safety:* Something must continue to *function* in order to keep the system safe.

Fault, Error, and Failure

A programmer may introduce a *fault* into a program by typing something unintended into an editor. A fault is a passive flaw.

Sometimes a fault may cause the program to perform in a way that produces an unintended outcome, the fault having caused an *error*.

Sometimes an error causes a *failure* of the system. In the words of EN 50128, "a failure has occurred if a functional unit is no longer able to perform its required function, i.e., a failure is an observable effect outside the system boundary arising from an internal error or fault. An error or fault does not always lead to a failure."

The important word in these descriptions is "sometimes." As the quotation from EN 50128 states, a fault does not always cause an error. If, for example, the programmer had meant to type

```
char x[10];
```

for a local variable, and accidentally typed

```
char x[11];
```

that is unlikely to cause an error unless the system is very short of memory. It is, however, a fault.

If a programmer created a system that opened a file every 10 seconds, but failed to close those files, then this *fault* would cause an *error* every 10 seconds (a leak of a file descriptor). However, that error would not cause a *failure* for some time — not for about 90 days for the computer on which this paragraph was typed, which permits up to 777,101 open file descriptors. For a real example, see Anecdote 9 on page 114.

As illustrated in Figure 2.1, an error at one level in a system may cause a fault at another. In that figure, a fault in Fred's training causes him to insert a bug in the code (he makes an error by inserting a fault). Fred loses his job over this mistake (a failure — at least as far as Fred is concerned). The fault in the code goes on to cause a failure in the overall system.

Differentiating between faults, errors, and failures allows us to tackle them in different ways — we want to reduce the number of faults inserted into the system, to detect and remove faults before they become errors, to detect errors and prevent them from becoming failures and, if a failure does occur, to handle it in the safest way possible.

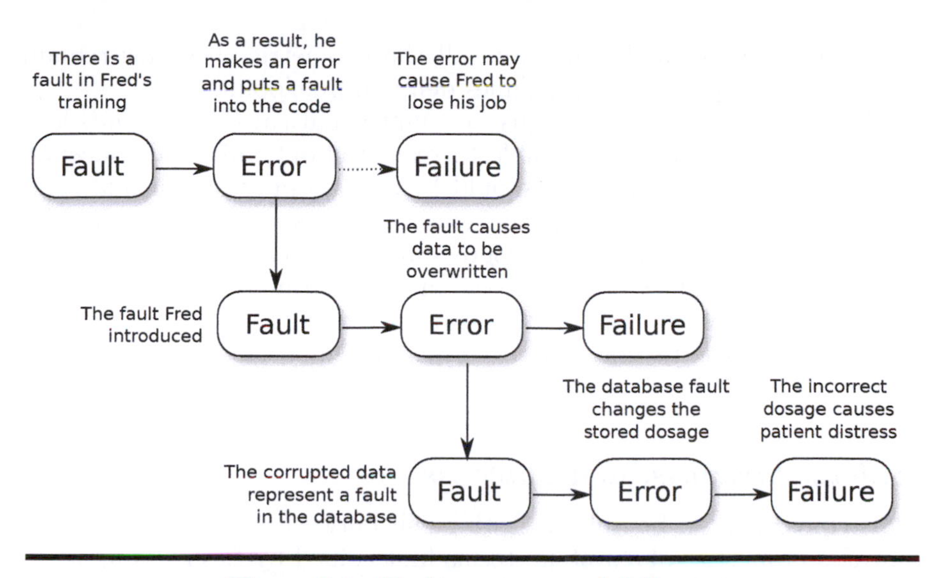

Figure 2.1 Faults, errors, and failures.

Formal, Semi-formal, and Informal Languages

Natural languages — English, Mandarin, German, Inuktitut — are ambiguous and subtle.

The oft-repeated reference given for an ex-employee, "You'll be lucky to get this person to work for you," illustrates the ambiguity of English. Another job reference, once apparently given for a candidate applying for a position as a philosophy professor, "This person has very nice handwriting," illustrates its subtlety.

Such a language is considered to be *informal* because both its syntax (grammar) and its semantics (meaning) are ambiguous.

It is possible to make a language's syntax formal, while retaining the flexibility of informal semantics and such languages are known as *semi-formal.* The universal modeling language, UML, and the systems modeling language, SysML, are both often used in a semi-formal manner. The syntax is well-defined in that an open, triangular arrow head always means "generalisation," and a stick figure always means "actor". However, when a block in a SysML diagram is labelled "Braking System," there is no definition of what that means, the everyday English usage being assumed.

The language of mathematics allows us to express ideas about universals. Mathematics doesn't understand what a triangle actually is, but it can prove that the angles in *any* triangle add up to 180 degrees. The principle of duality in geometry emphasizes the abstraction from real objects by saying that, with a few constraints, we can exchange the terms "point" and "line" and all the geometric theorems remain true. For example, two points/lines define a line/point.

If we can express our designs in a mathematical (i.e., *formal)* language, then it should be possible to reason about universal concepts. Given our system design, we should be able to prove mathematically that for *any* combination and sequence of requests and interrupts, the system will continue to make progress and not lock up.

Proving the correctness of a system differs from testing it: *Testing* demonstrates that it works for the particular set of conditions that occurred during the testing, whereas *proving* demonstrates that it works under all conditions.

Safety, Liveness, and Fairness

It is convenient to be able to distinguish between safety, liveness and fairness properties of a system, terms first coined by Leslie Lamport in 1983. Informally:

A safety property says that the system never does anything bad, i.e., it remains safe. For example, the system will never transmit message A unless it has previously received message B.

A liveness property says that the system eventually does something good, i.e., it will make progress and not live- or dead-lock. For example, if message B is received, then message A will always eventually be transmitted.

A fairness property says that, if the system attempts actions, then those actions will eventually succeed, i.e., the system will not allow one activity to "take over" so that another activity's requests remain permanently unanswered.

There are many variations on fairness ranging from strong fairness (if the system attempts X infinitely often, then it will succeed infinitely often) to weaker forms (e.g., if the system continuously attempts X, then it will succeed at least once).

In the case of a safety property, there is a finite (although possibly very large) set of states to be examined to check whether that property is valid. In this case the check may be practically infeasible, but can theoretically be performed. For a liveness property, the number of possible states that need to be checked is potentially infinite because of the term "eventually."

Backward and Forward Error Recovery

When a fault causes an error, that error can sometimes be trapped and some form of recovery can take place before a failure occurs. The two main forms of error recovery are:

Backward Error Recovery.
Once an error is discovered, the system returns to a previously-stored state that is known (or believed) to be consistent and sane. Whether the input that apparently caused the error is then reapplied or discarded, and how the client is notified, vary between implementations.

One of the disadvantages of backward error recovery is the need to store the system's state information before handling an incoming request so that it is available if rollback is required.

Forward Error Recovery.
This removes the main disadvantage of backward error recovery — the need to save the system's state — by moving to a predefined sane state independent of both the previous state and the input that caused the error.

Accidental Systems

In the presentation associated with reference [3], Martyn Thomas described an experiment wherein global positioning system (GPS) signals were deliberately jammed in part of the North Sea and a ship was taken through the area to see what would happen. As expected, the position of the ship became imprecise (sometimes the ship was in the middle of Norway, sometimes in the middle of Germany!), but unexpectedly, the ship's radar also stopped working. The experimental team contacted the company that manufactured the radar and was told that GPS was not used within it. They then contacted the manufacturers of the components in the radar and found that one small component was using GPS timing signals.

The radar constitutes an "accidental system." The manufacturer of the small component knew that any system into which it was incorporated would rely on the integrity of the GPS signals, but was not aware of the radar system into which the component was being incorporated. The designer of the radar was unaware of the dependency. No one had the full picture, and a system had been built with accidental dependencies. The system would not have been tested in the face of GPS jamming, because the reliance on GPS was unknown.

Embedded software-based systems can also become accidental systems. Perhaps the designers of the complete system were unaware that an incorporated component was single-threaded or included recursion that could exceed the allocated stack size.

Software-Specific Terminology

Software

In a book dedicated to embedded software, it may seem strange to ask what *software* is — intuitively people feel that they can differentiate between software and hardware. If it hurts when dropped on one's foot, it's hardware.

But what about an Application-Specific Integrated Circuit (ASIC) designed by a software tool? What about 5000 lines of C code generated by a modeling tool?

Reference [4], published by the Rail Safety and Standards Board, treats the distinction between hardware and software pragmatically by noting that software is intrinsically more complex than hardware:

> *As a rule of thumb, we suggest that if a device has few enough internal stored states that it is practical to cover them all in testing, it may be better to regard it as hardware and to show that it meets its safety requirements by analysis of the design and testing of the completed device, including exhaustive testing of all input and state combinations.*

This is the definition that we will use in this book. If the system is sufficiently simple that it can be tested exhaustively, then it will be treated as hardware; otherwise as software.

Note that this is a one-way definition. Certainly, the embedded systems considered in this book are too complex for exhaustive testing and will therefore be treated as software. However, it has also been impossible to test some hardware devices exhaustively for many years. For example, even a 64 kbit RAM (ludicrously small by today's standards) has $2^{64 \times 1024}$ possible states — a number containing 19,728 digits — and, under the definition in reference [4], could be considered software. I will ignore that interpretation.

There is another aspect of software that is often neglected: configuration data. Commission Regulation (EC) No 482/2008 of the European Union, explicitly defines the term "software" to include its associated configuration data:

> *'software' means computer programmes and corresponding configuration data . . .*
> *'configuration data' means data that configures a generic software system to a particular instance of its use;*

Following the crash of an Airbus A400M military transport aircraft on 9th May 2015, an executive of Airbus Group confirmed in an interview with the German newspaper *Handelsblatt* that the crash was caused by a faulty software configuration. In particular, the executive said that "The error was not in the code itself, but in configuration settings programmed into the electronic control unit (ECU) of the engines."

Validating a program together with its configuration data is not well understood because of the combinatorial problem associated with configuration. If a program takes 5 configuration options, each of which can have any of 3 values, then there are really $3^5 = 243$ different programs to verify — see the discussion of combinatorial testing on page 257. The importance of including configuration data in program verification is being increasingly recognized, and new standards are likely to emerge over the next few years.

Heisenbugs and Bohrbugs

Niels Bohr described a solid, billiard-ball model of the atom and won the 1922 Nobel Prize for his work. Werner Heisenberg, a student and colleague of Bohr's, enhanced this model to make the atom much more uncertain: a probabilistic model of where the electrons, neutrons and protons might be. This work won Heisenberg the Nobel Prize in 1932.

The etymology of the related terms *Bohrbug* and *Heisenbug* is unclear, although Jim Gray used the term "Heisenbug" in a 1983 paper. It is believed to have originated sometime in the early 1980s.

As with the atoms, so with the bugs. A Bohrbug is a nicely defined, solid bug with properties that don't change when debug code is added to find the bug. Every time a particular line of code is executed with particular values of the input variables, the system misbehaves.

A Heisenbug, in contrast, is a will-of-the-wisp that appears and disappears in a manner that makes it extremely elusive. Heisenbugs are caused by subtle timing problems: Thread A running on processor core 0 does something that affects thread B running on processor core 1 if, and only if, an interrupt of type C occurs, ...

Heisenbugs are reported by test groups as being non-reproducible: "The eighth time that test case 1243 was run, the system crashed. We've run that test case several dozen times since, and no other failure has occurred. No memory dump is available. Please fix the bug."

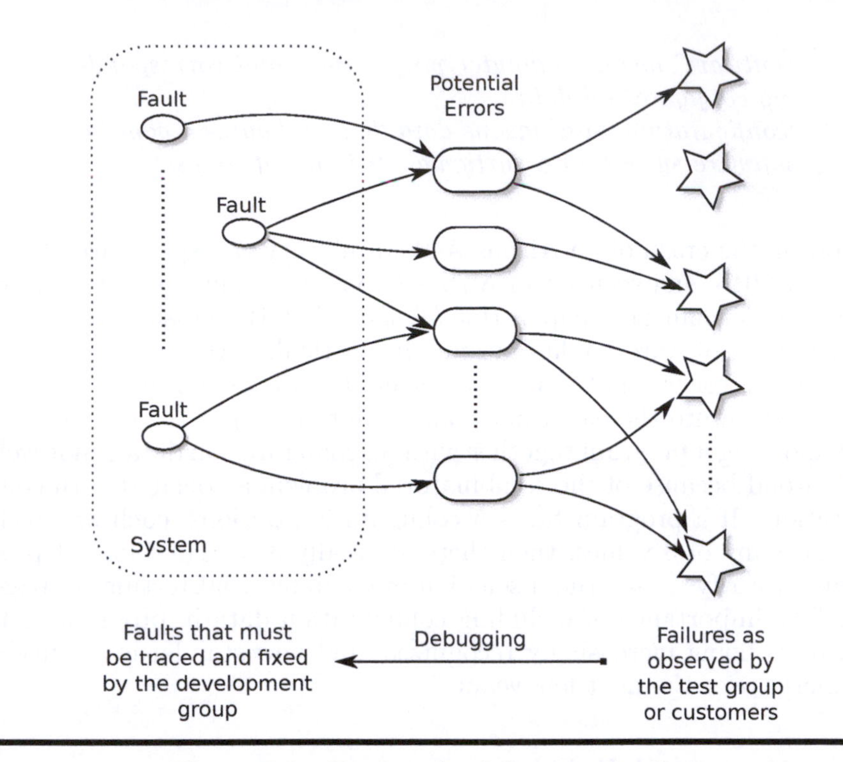

Figure 2.2 Tracing the failure to the fault.

It is a characteristic of Heisenbugs that the same fault can give rise to different failures, and the failures can occur much later than when the fault was invoked (and in an entirely different piece of code). This is what makes debugging difficult: Testing detects failures, whereas the developers must fix faults. For all faults, but particularly for Heisenbugs, the path from the fault to the failure is many-to-many and this makes tracing the fault, given the failure, very difficult — see Figure 2.2.

At the level that can be observed by a tester, many failures appear to be identical. The errors that cause the failures are invisible to both tester and programmer, and different faults may give rise to essentially identical errors.

Jim Gray made the point, although not in the context of safety-critical systems, that users prefer Heisenbugs to Bohrbugs; they prefer a failure occurring randomly about once a month to a failure every time a certain command is entered in a particular context. For this reason, research has been carried out (e.g., reference [5]) on converting Bohrbugs into Heisenbugs. For example, if more memory is allocated for an object than is strictly required and the memory is allocated probabilistically, a memory overflow becomes a Heisenbug rather than a Bohrbug.

The elusive nature of Heisenbugs make them particularly difficult to fix, and some researchers, particularly Andrea Borr (see, for example, reference [6]), have found that attempts to fix Heisenbugs generally make the situation worse than it was before.

Irreverent programmers have extended the Bohrbug and Heisenbug terminology in more comical directions. Named after Erwin Schrödinger (Nobel Laureate in 1933), the Schrödingbug, for example, is a fault that has been in the code for many years and has never caused a problem in the field. While reading the code one day, a programmer notices the bug. Immediately, field reports start to flow in as the bug, having now been observed, like Schrödinger's cat suddenly wakes up in the field.

Real-Time

The term *real-time* is perhaps one of the most confusing and misused terms in software development. A system is real-time simply if its required behavior includes a time constraint.

Hence, a program that calculates a weather forecast from currently-available data is a real-time program because an excellent forecast for a period 12 hours ahead is useless if it takes 14 hours to produce. Similarly, the payroll calculations performed by companies to pay their employees are real-time because failure to complete them on time could lead to very unhappy employees.

Within the embedded world, a system is real-time if part of its contract specifies a time within which it will respond: "This system reads the sensor every 10 ± 2ms and makes the processed data available within 5 ms of the reading."

Fallacy 1 *"Real-time" means "fast."*

Real-time systems tend to be slower than non-real-time ones because of the time buffering needed to ensure that deadlines are always met, irrespective of the timing of other subsystems.

Programming by Contract

The concept of programming by contract or contract-first development was devised by Bertrand Meyer for the Eiffel programming language in the late 1980s* and is now available in other languages, such as D. Programming by contract defines a three-part formal contract between a client and a server:

1. The client has an obligation to ensure that its request meets the server's preconditions. For example, a precondition for invoking the function to add an item to a queue may be that the queue is not already full.
2. The server guarantees that certain post-conditions will apply to its response. In the case of adding an item to a queue, this would include the fact that the queue has exactly one more element than it had previously and that the last element on the queue is the one provided by the client.
3. The server guarantees that certain internal invariants will apply on exit so that the next call to it will also be answered correctly.

If the contracts are coded into the server program when it is written, then not only can they be checked at runtime to catch otherwise difficult to find bugs, but they can also be used to assist static analysis tools with their work: see reference [7] and Chapter 18.

For example, given the queue `add()` function above, a static analysis tool could analyze the whole code base to determine whether it could ever be possible for the function to be called when the queue was already full.

References

1. C. Babbage, "On the Mathematical Powers of the Calculating Engine." (Published in Origins of Digital Computers: Selected Papers (ed. B. Randell) Springer-Verlag, 1973), 1837.
2. M. P. E. Heimdahl, "Safety and software intensive systems: Challenges old and new," in *2007 Future of Software Engineering*, FOSE '07, (Washington, DC, USA), pp. 137–152, IEEE Computer Society, 2007.
3. M. Thomas, "Accidental Systems, Hidden Assumptions and Safety Assurance," in *SSS* (C. Dale and T. Anderson, eds.), pp. 1–9, Springer, 2012.

* Note that Eiffel Software is the owner of the copyright on the related term *Design by Contract.*

4. Rail Safety and Standards Board, "Engineering Safety Management (The Yellow Book)," 2007. Available from `www.yellowbook-rail.org.uk`.
5. E. D. Berger, "Software needs seatbelts and airbags," *Commun. ACM*, vol. 55, pp. 48–53, Sept. 2012.
6. A. J. Borr and C. Wilhelmy, "Highly-Available Data Services for UNIX Client-Server Networks: Why Fault Tolerant Hardware Isn't the Answer," in *Hardware and Software Architectures for Fault Tolerance*, pp. 285–304, 1993.
7. R. Ceballos, R. M. Gasca, and D. Borrego, "Constraint Satisfaction Techniques for Diagnosing Errors in Design by Contract Software," in *Proceedings of the 2005 conference on specification and verification of component-based systems*, SAVCBS '05, (New York, NY, USA), ACM, 2005.

Chapter 3

Safety Standards and Certification

> *All too often, writers of standards focus on questions of what constitutes good practice, and lose sight of what the followers of those standards truly need to demonstrate in order to show safety. Safety is demonstrated not by compliance with prescribed processes, but by assessing hazards, mitigating those hazards, and showing that the residual risk is acceptable.*
>
> From reference [1]

Standards Bodies

It is said that the good thing about standards is that, if you don't like one, then there is always another that you can apply instead. In reference [2], John McDermid and Andrew Rae actually say "it is hard to count [the safety engineering standards], but there are probably several hundred." And, of course, there is no clear boundary around the safety standards — is IEC 29119, a standard that covers software testing, a safety standard or not?

There are also numerous bodies that produce standards. In this book I concentrate on a small number of standards produced by two organizations: the International Electrotechnical Commission (IEC) and the International Organization for Standards (ISO).

Both of these organizations are based in Switzerland, and although each produces its own standards, many standards are cross-issued. Thus, ISO/IEC/IEEE 42010, the standard describing how the architecture of a system may be captured, has joint prefixes.

One characteristic of both the IEC and ISO is that, unlike standards bodies such as the Internet Engineering Task Force (IETF) and the Distributed Management Task Force (DMTF), they charge for access to the standards. A single-user licence to read IEC 61508 costs around $1600 US: about $2.39 US for each of the 670 pages in the English edition. Note that this is for a single-user licence and if several people in an organization need to be able to read it, then either several licences or a more expensive site licence must be bought.

Anecdote 2 *One company for which I worked purchased a copy of one of the safety standards for me — a single reader licence. Unfortunately, when the PDF version was downloaded, it was marked on every page with the name of the person within the company's purchasing department who had actually placed the order. Strictly, only that person was allowed to read it, and I was not allowed to look at it.*

This revenue source has been criticized for putting a barrier between companies, particularly small companies, and legal copies of important standards. It has been suggested that the IEC and ISO should provide the standards free of charge and get their revenue from the standardization process itself: a royalty whenever a certification body issues a certificate against a standard. This would encourage the production of useful standards, as the more a standard was used, the greater would be the revenue from it. However, questions of conflict of interest might arise if this model were adopted, because the IEC and ISO are required to be independent of the certification process.

Standards from the IEC and ISO are produced in basically the same way: Industry experts work to create a standard, and the acceptance of the standard depends on the results of a vote, where each country participating in the IEC or ISO gets a single vote. Thus, the UK, Germany, China, and the USA each get one vote, as do Luxembourg, Mongolia, and Bhutan.

To influence the vote on a particular ISO standard, an engineer needs to become associated with her country's standards body: the Bureau voor Normalisatie in Belgium, the British Standards Institution in the UK, the Deutsches Institut für Normung (DIN) in Germany, the American National Standards Institute in the USA, the Standards Council of Canada, etc.

Accreditation and Certification

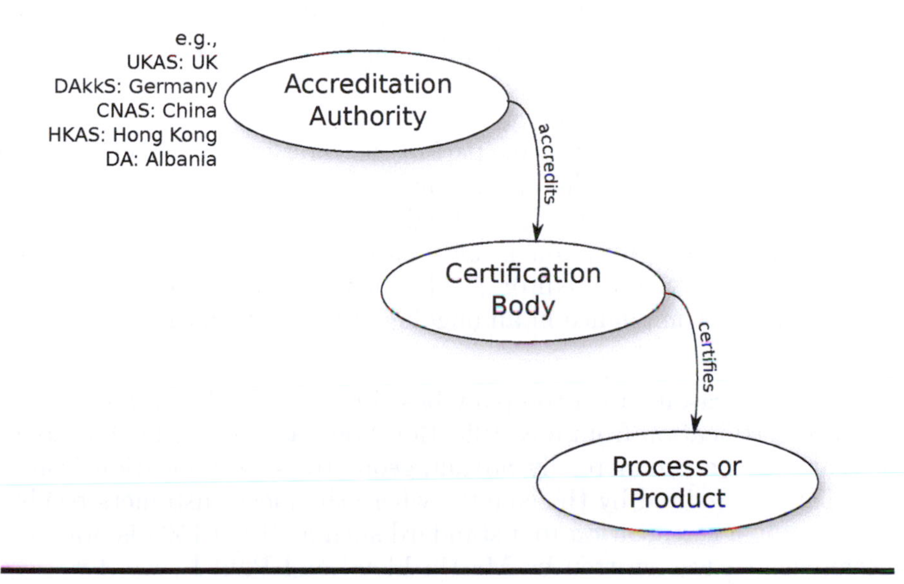

Figure 3.1 Accreditation and certification.

Figure 3.1 illustrates the accreditation and certification process:

Certifying a product or process.
A company that wants to have one of its processes or products certified against a particular standard could self-declare that the process or product meets all the requirements of the appropriate standard, possibly employing an external auditing company to confirm the declaration.

Alternatively, the company could employ an independent certification body to certify the process or product.

Accreditation of certification bodies.
In principle, any company could act as an external certification body, but companies can become "accredited" to issue certificates for a particular standard. The certificate from a body that is not accredited is less commercially valuable than a certificate

from an accredited body.

Certification bodies are accredited by national accreditation authorities to carry out particular certifications. Thus, a certification body may be accredited to carry out certifications of hardware (but not software) against IEC 61508 up to safety integrity level 2 (SIL 2).

A company looking for a certification body should carefully check the accreditations of the candidates to ensure that the certificate that will be finally issued will be credible and commercially valuable.

Accrediting the accreditors.

Each country that forms part of the International Accreditation Forum (IAF)*, has one or more accreditation body members that accredit certification bodies. As there is no higher body, the obvious question is *quis custodiet ipsos custodes?*— who accredits the accreditors? The answer is that they verify each other's compliance in an ongoing cycle of evaluations.

There is no reason why a company based in a particular country needs to seek certification from a certification body accredited in that country; it may be commercially advantageous to get certification from a company accredited by the country where the most customers reside.

Getting a certification to a standard such as IEC 61508 is not easy. Reference [3] was written by Martin Lloyd and Paul Reeve, two very experienced auditors working at that time for a UK certification body. The reference describes the success rate that they had recently seen for IEC 61508 and IEC 61511 certifications of software-based products. Of the twelve companies that had completed the certification process at the time of the report, only three (25%) had achieved certification — the others had failed, in some cases after investing a lot of money and time. Reference [3] is particularly useful because it lists some of the common reasons for failure. Martin Lloyd updated the paper in early 2015 (reference [4]) to include experience between 2010 and 2015. Many of the same reasons for failure still exist, but the failure rate has reduced. Martin believes that this is because companies are now more aware of what is required and are less likely to start the certification process if they do not have the necessary evidence.

To maximize the chances of getting a product certified, it is useful to engage the certification company early in the development process. In particular, it can be useful to discuss the structure of the argument that will be presented in the safety case (see page 61) with the auditors

* `http://www.iaf.nu`

from the certification company well before the final audit. Even before the evidence is added, a confirmation or even an agreement that the argument will be satisfactory if backed by evidence is well worth having.

Why Do We Need These Standards?

In some areas, the need for standards is very clear. If there were no standard defining the shape and size of the electrical sockets in our houses, then it would be impossible to buy an electrical device with confidence that it would connect to the electrical supply. Indeed, it would be very convenient if electrical sockets were standardized across the world.

The need for standards defining how devices with safety-critical requirements are to be built and classified is less obvious.

Historically, the creation of standards (and not just safety-related ones) has often been driven by disasters. A disaster occurs and the public demands that it never occur again. The response is the creation of a standard to which industry must comply, with the intent of raising the quality of products and reducing the chance of a repetition. In this sense, the standard provides protection for a public that does not understand the engineering process.

From the point of view of a product development organization, standards can be a useful definition of "adequate" practices. A company new to the development of a safety-critical application would find a useful set of tools and techniques listed in IEC 61508, ISO 26262, EN 50128, or other standards. Complying with such standards could also be a mitigation in the event of a court case: "We were following industry best practices."

From the point of view of a company buying a component or subsystem for use in a safety-critical application, the standard forms a shorthand that simplifies contractual relationships: "The component must comply with IEC 61508 at SIL3" is easier to write than an explicit statement of the acceptable failure rate, development processes, etc.

During a roundtable discussion at a conference of safety engineers I attended recently, the panel was asked whether system safety had improved over the last decade as a result of the number of standards that had appeared. The unanimous agreement of the experts was that safety has improved, but because of more attention being paid to software development, rather than because of the application of the standards.

Goal- and Prescription-Based Standards

Standards can be classified as lying on a spectrum between two extremes: prescriptive and goal-based.

Prescriptive standards state what must, and must not, be done during the product development. They prescribe and proscribe processes, procedures, techniques, tools, etc. They don't state the goal of the development, just the means of achieving it.

If a certificate were issued against a prescriptive standard, a company would typically not claim that, "Our product is safe." Rather, it would claim that, "Our product meets the requirements of IEC xxxxx."

Goal-based standards, in contrast, state the goal that must be achieved and leave the selection of appropriate processes, procedures, techniques, and tools to the development organization. This places a greater burden on the development team: not only need it define its processes, tools, and techniques, it must also justify their adequacy. The work of auditing for certification is also much more difficult because the auditor cannot prepare a checklist from the standard and simply tick things off.

If a certificate were issued against a goal-based standard, a company could, in principle, claim that, "Our product is safe."

The argument in favor of goal-based standards is that they don't lock in obsolete technologies. The cycle for updating a standard typically lasts for a decade or more, and software techniques are changing much more rapidly than that. Many tools used today could only have been dreamed of 10 years ago. An out-of-date prescriptive standard may lock in the use of an out-of-date technique or tool.

In the first edition of IEC 61508, published in 2000, almost all of the references to tools and techniques came from the 1980s, with some from the 1970s. It is dispiriting and wasteful for a company to apply techniques to a development only because the standard demands it: "We know that this is useless, but the standard says we must do it, and it's easier to do it rather than argue about it with our auditor." This is a particularly plaintive cry when the assigned auditor has read the standard thoroughly, but may not be aware of the rapidly changing nature of the software industry.

On the spectrum between purely goal-based and purely prescriptive standards, it should be said that most of those discussed below fall closer to the prescriptive end. The avionics standard, DO-178C, is perhaps the most goal-based that we currently have.

Functional Safety Standards

The concept of functional safety is introduced on page 15; this section describes a few of the standards relating to it.

IEC 61508

The Standard

IEC 61508 is perhaps the most fundamental standard relating to functional safety; see Figure 3.2. This standard is entitled *Functional safety of electrical/electronic/programmable electronic safety-related systems* and comes in seven parts, the first three of which are normative,* covering general project management and the development of hardware and software components.

The remaining four sections are informative and contain definitions of the vocabulary and examples and guidelines for applying the first three parts.

IEC 61508 was first issued around the year 2000 (the parts were issued at slightly different times) and the second edition was published in 2010. For those moving from issue 1 to issue 2, the IEC provides (on receipt of payment) a CD with all of the changes and the justifications for the changes marked.

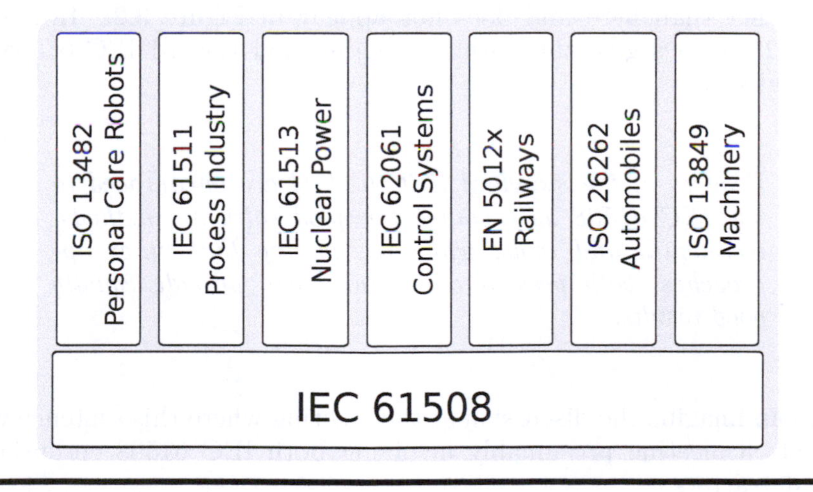

Figure 3.2 Some functional safety standards.

* A "normative" part of a standard defines what must, or must not, be done. Most standards also include informative sections that provide explanation, guidance, and additional information, but that do not form part of the standard itself.

The generic model used in IEC 61508 is of a piece of industrial equipment that can inherently give rise to safety hazards, and a safety function that monitors it and moves it into a safe state when a hazardous situation is detected.

From a software development point of view, IEC 61508's most valuable contributions are the 18 tables in Annexes A and B of part 3, each of which lists tools and techniques that are specifically *not recommended*, *recommended*, or *highly recommended* for each stage of the software development life cycle.

Specializations of IEC 61508

As can be seen from Figure 3.2, IEC 61508 has been specialized for a number of industries. The level of linkage between IEC 61508 and the industry-specific standards varies between industries. EN 5012x, the railway standards, rely explicitly on IEC 61508 and include many tables copied from it with little modification.* The automotive standard, ISO 26262, also copies many of the tables from IEC 61508, making subtle and, in places, not so subtle changes. Other standards, such as ISO 13482, are based much less directly on IEC 61508, although, even there, ideas derived from IEC 61508 are included.

Perhaps the most interesting reliance on IEC 61508 is included in the medical specification IEC 62304. This specification is not directly concerned with functional safety and is therefore described below (page 41) rather than here and does not appear in Figure 3.2. However, IEC 62304 does give this rather grudging nod toward IEC 61508 in Appendix C.7:

> *Readers of the standard [IEC 62304] are encouraged to use IEC 61508 as a source for good software methods, techniques and tools, while recognizing that other approaches, both present and future, can provide equally good results.*

One can imagine the discussion at the meeting where this sentence was added: a meeting presumably involving both IEC 61508 enthusiasts and detractors.

* I use the term "EN 5012x" to include EN 50126, EN 50128 and EN 50129, all of which are expected to be combined into one standard in 2016 or 2017.

Safety and Failure

IEC 61508 and its derivatives are firmly rooted in the idea that the study of device safety is the study of device failure. This was the unchallenged belief throughout the 20th century, starting in the early days of safety engineering addressing the question of whether a two-engined or four-engined aircraft was safer.**

Superficially, the link between safety and failure seems strong: the safety function must continue to operate to provide safety, and so its failure represents a threat to safety. This approach has been challenged by Nancy Leveson as being too simplistic for 21st century systems. She argues (see, for example, reference [5]) that systems today are so complex that it is impossible to predict all of the possible interactions that might lead to failure, and we need more sophisticated models to assess a system's safety. Leveson has proposed an approach called "system-theoretic accident model and processes" (STAMP), which is based on systems theory rather than reliability theory. Figure 3.3 illustrates the fundamental abstraction of STAMP. A system consists of a hierarchy of controlled processes, and each controller has a "mental" model of the process it is controlling.

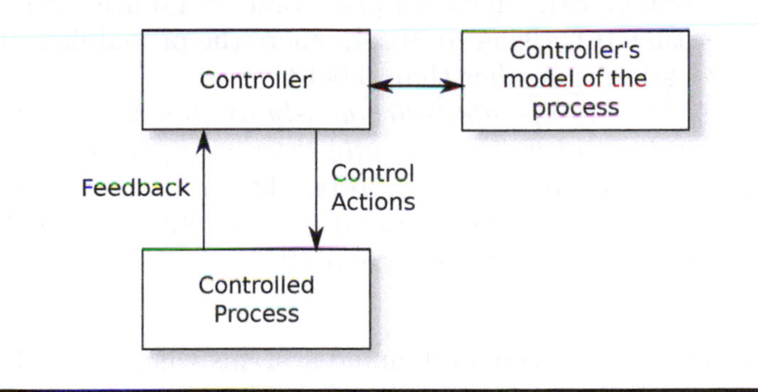

Figure 3.3 The STAMP model.

Dangerous situations occur when control actions do not enforce safety, when control actions are incorrectly co-ordinated between controllers, and when the control structure gets out of line with the process being controlled, causing the controller to issue an inappropriate control action. In the STAMP model, the controller may be a human or a software or hardware subsystem. Because the control loop shown in

** A four-engined aircraft faces twice as many engine failures as a two-engined one, but each failure is presumably less severe.

Figure 3.3 is itself part of a larger system, any incorrect action on its part can lead to inconsistency at a higher level.

IEC 61508 and SILs

In contrast with the STAMP model, IEC 61508 defines a set of *safety integrity levels* (SILs) based purely on failure probability, thereby binding the safety and failure concepts together. Two types of safety system are considered:

Systems that are used only when a particular demand arises.
These are known as on-demand systems and can be contrasted with systems that must work continuously to maintain safety. IEC 61508 divides on-demand systems into those that are expected to be invoked infrequently (less often than once per year) and those that would be invoked more often.

As an example, consider the final defense against a meltdown of a nuclear reactor should all other defenses fail: It is intended that the system *never* be invoked; certainly, it would not be expected to have to operate more often than once per year. For such a system, SIL 1 means a probability of failure when invoked smaller than 0.1, rising to SIL 4, where the probability of failure on invocation is smaller than 0.0001.

Systems that have to operate continuously to maintain safety or operate on-demand more frequently than once per year.
Here, SIL 1 means having a probability of dangerous failure of less than 10^{-5} per hour of operation, stretching to SIL 4, where the probability must be less than 10^{-8}.

Note the distinction between on-demand systems where the SIL defines the probability of failure on demand and continuous systems where the SIL defines the probability of failure per hour of use.

In order to demonstrate compliance with, for example, SIL 4 in a continuously operating system, it is necessary to justify a failure rate of less than 10^{-8} per hour of operation. As Odd Nordland has pointed out (reference [6]), these numbers are statistical, and to be able to argue about them for a particular device, it is necessary to have a statistical sample of failures for analysis. Table 3.1 lists the probability of a failure in a given time for a system that meets SIL 4, and it can be seen that waiting for a statistical sample (say, 10 or 100) of failures would need a lot of patience. It can be argued that, apart from some specialized systems, SIL 4 is not only an unreachable target, but is a target that could never be justified.

Table 3.1 Probability of failure for a SIL 4 device.

Hours	Years	Probability of a failure
100,000	11	0.001
1,000,000	114	0.00995
10,000,000	1,141	0.09516
100,000,000	11,408	0.63212
1,000,000,000	114,077	0.99995

One exception might be an embedded component that is used in many devices and for which good records of usage and failures are kept. There are, for example, about 250 million cars in the USA. If a particular component were installed in 10% of those and each car were in operation for an average of 2 hours per day, then the component would achieve about 2×10^{10} hours of operation per year. However, in order to make use of these hours of operation, it would be necessary to know the numbers of failures, and these might be difficult to obtain. If the software in the lane change warning in a car fails, the driver would probably turn it off and on again and never bother to report the failure.

ISO 26262

The Standard

As illustrated in Figure 3.2, ISO 26262 is the specialization of IEC 61508 for production cars. The introductory text in the standard specifically excludes speciality cars and any form of buses or trucks: the electronic systems in these may require development in accordance with international standards, but ISO 26262 is not that standard.

ISO 26262 is in ten parts, the first nine parts having been issued in 2011, and part 10, the guideline for applying the other nine, appearing in 2012.

Parts 2 to 7 describe the approved process for product development: moving through safety management (part 2), the product's concept phase (part 3), system-level development (part 4), hardware development (part 5), software development (part 6), and production (part 7). The other parts provide supporting material for these.

Part 6 of ISO 26262 deals with software, and, like IEC 61508, includes tables of recommended and highly-recommended techniques for each automotive safety integrity level — ASIL. There are 16 such tables and their contents overlap those in IEC 61508, but are by no means identical.

ISO 26262 and ASILs

ISO 26262 replaces the concept of a SIL with the concept of ASIL. Unlike IEC 61508's SIL, the ASIL is not based on probability of failure per hour of use. Instead, the events that could cause injury are listed and an ASIL is calculated for each of them by considering three aspects:

What type of injuries could result from this event?
: Annex B.2 of part 3 of ISO 26262 presents medical descriptions of possible injuries, for example "fractures of the cervical vertebræ above the third cervical vertebra with damage to the spinal cord." These descriptions can be used to assess the severity of the risk on a scale from S0 (no injuries) to S3 (life-threatening and fatal injuries).

  Take, as an example, an automated mechanism to wind a car's rear window up or down. If this were to close the window while a passenger had her hand in the way and it did not detect the resistance and stop the window, it could cause minor injury (S1 — light and moderate injuries). However, another event might be that it closed on a child with his head out of the window. That could cause more serious injury (perhaps S2 — severe and life-threatening injuries where survival is probable).

How likely is the event to occur in normal operation?
: ISO 26262 assesses the likelihood on a scale from E0 (incredible) to E4 (high probability) and gives further guidance that E1 is applied to events that occur less often than once a year and E2 to events that occur a few times per year. Notice that this assessment is dependent on the geographical area where the car is being sold. For example, anti-lock brakes are activated much more often in a Canadian winter than in southern California.

  I have not done the necessary research, but would suspect that winding a window up would be something that would occur perhaps a few times each year — E2.

How many drivers could control the situation and avoid injury?
: The scale here is from C0 (controllable in general) to C3 (difficult to control or uncontrollable). Again, research would be needed, but I suspect that the chances that the driver or other passenger could control the window closing on the hand or neck is fairly low — it would probably happen before anyone noticed. So this would probably be classified as C3.

Given these three classifications, the ASIL associated with the event can be read from a table; see Table 3.2.

Table 3.2 Assignment of ASIL

Severity	Likelihood	Controllability		
		C1	C2	C3
S1	E1	QM	QM	QM
	E2	QM	QM	QM
	E3	QM	QM	ASIL A
	E4	QM	ASIL A	ASIL B
S2	E1	QM	QM	QM
	E2	QM	QM	ASIL A
	E3	QM	ASIL A	ASIL B
	E4	ASIL A	ASIL B	ASIL C
S3	E1	QM	QM	ASIL A
	E2	QM	ASIL A	ASIL B
	E3	ASIL A	ASIL B	ASIL C
	E4	ASIL B	ASIL C	ASIL D

In the example of the window closing on a child's neck, the classifications were S2, E2, and C3. Consulting Table 3.2 it can be seen that the safety function that prevents this injury would need to be assessed against the requirements for ASIL A.

Some other observations can be made from Table 3.2. In particular, there are many positions for which no ASIL is assigned, "QM" (quality management) being listed instead. For these combinations, there is no requirement to comply with ISO 26262, but the component must be developed in accordance with a quality management process that has been approved to an international standard such as ISO 16949 ("Quality management systems — Particular requirements for the application of ISO 9001:2008 for automotive production and relevant service part organizations") or ISO 9001 ("Quality Management").

Also, it can be seen from Table 3.2 that ASIL D is very rare — it occurs in only one position in the table, the position combining highest severity of injury with highest probability of occurring with the lowest possibility of the situation being controlled. There has been an ASIL inflation over the years, with many companies asking for ASIL D components unnecessarily.

Mixed-ASIL Systems

Apart from the smallest components, it would be rare to find a complete system with a single ASIL.

In the past, there has generally been a clear division between the safety-related aspects of a car and the entertainment, navigation, and other functions that do not affect safety. The safety-related applications have been developed in isolation both from the nonsafety ones and from each other, in accordance with a series of specifications known as AutoSAR (AUTomotive Open System ARchitecture) — see, for example, reference [7], which specifies the requirements on an AutoSAR operating system.

Today, such physical isolation is no longer possible, as manufacturers are looking to reduce costs by combining hardware components. A single, multicore processor can now handle the work previously handled by a dozen or more micro-controllers, and sensors can be shared — a technique known as sensor fusion. If two subsystems each need a camera pointing backward from essentially the same place, why should it be necessary to install two cameras?

As soon as two subsystems, possibly of different ASILs, share hardware the question of isolation arises: How can the design guarantee that one subsystem will not interfere with another?

Annex D of ISO 26262-6 and Annex F of IEC 61508-3 address the question of isolation between components and list some potential sources of unwanted interaction that need to be addressed during the design:

Timing.
: How can we guarantee that one subsystem will not absorb so much processor time that another subsystem is starved?

Deadlocks and livelocks.
: What prevents two unrelated subsystems that share a peripheral (e.g., a mutex) from forming a deadlock or livelock?

Memory.
: How does the design guarantee that one subsystem will not write over the memory used by another subsystem?

Messaging.
: If one subsystem is relying on receiving messages from another subsystem then how does the system guarantee continued safety if a message is lost, corrupted, delayed, or duplicated?

Although this is a useful list, it is far from complete. Most modern operating systems provide the tools to ensure memory protection, fair sharing of processor time, etc. More insidious are the interactions between components that lie outside the control of both the subsystems and the operating system. On many processors "hidden" elements, such as instruction and data caches, graphics coprocessors, DMA controllers and NVRAM controllers, share resources (buses, etc.) with the applications in a way that is invisible and uncontrollable by the applications. How is the necessary isolation then guaranteed?

ISO 26262 and SEooCs

ISO 26262 introduces the concept of a "safety element out of context" (SEooC) and the whole of Section 9 of Part 10 is dedicated to describing SEooCs. A SEooC is an element* that is to be used in an automotive system, but which was not specifically designed for use in that system.

An automotive system might, for example, need a database and, rather than design and develop a new one, the designers might instead use MySQL. In this case, MySQL would be a SEooC.

When using a SEooC, the designer needs to state clearly what is expected of the element and then ensure that the SEooC meets those requirements, either "out of the box" or after modification. If the SEooC has been developed by a third party, then, once it has been selected, the system designer must monitor the bug reports for the SEooC and carry out an impact analysis to determine whether any bug could affect the safety of the system into which it has been included. This can be an onerous task — the bug lists for software such as the Linux kernel are long and dynamic.

Note that the term SOUP (software of unknown provenance) is used in IEC 62304, the medical device standard, in much the same way as SEooC is used in ISO 26262.

IEC 62304 and ISO 14971

Many standards relate to safety issues (see Figure 3.2 for a few of them), but I think that it is worth including one that is not related to *functional* safety: IEC 62304. This standard was published in 2006 to address the life cycle processes for software in medical devices.

In a manner similar to IEC 61508's definition of SILs and ISO 26262's ASILs, IEC 62304 defines three classes of risk that might be associated with a particular medical device. Devices where no injury or damage to the health of a patient, device operator, or onlooker would result if the software failed to behave as specified are classified as Class A. If non-serious injury is possible, then the device is Class B. If death or serious injury is possible, then the device is Class C.

* The terms "element", "system" and "item" have precise meanings in ISO 26262 — I use them here generically.

This classification is defined in section 4.3 of IEC 62304 and is followed by a statement that has generated much debate: If a dangerous condition can arise"from a failure of the software system to behave as specified, the probability of such failure shall be assumed to be 100 percent." This is confusing, because no time frame is given, and many designers with whom I have worked have asked how such an assumption can be incorporated into a failure model.

Anecdote 3 *The IEC 62304 statement that it must be assumed that software always fails is the opposite of the situation I once met on a railway project, where the safety engineer was adamant that the probability of failure of the software should be set to zero because "software never wears out and therefore never fails." I pointed out that this flew in the face of common experience, but could not budge him.*

I am not a lawyer and cannot interpret the 100% comment in IEC 62304 other than by ensuring that, when the failure model for the medical device is created (see page 67), the failure rates of its software components are greater than zero. This implies that the software will fail with 100% probability (as required by IEC 62304), albeit over an infinite time (which was probably not what was meant by IEC 62304). I suspect that the sentence was added to the standard to avoid the type of discussion illustrated in Anecdote 3.

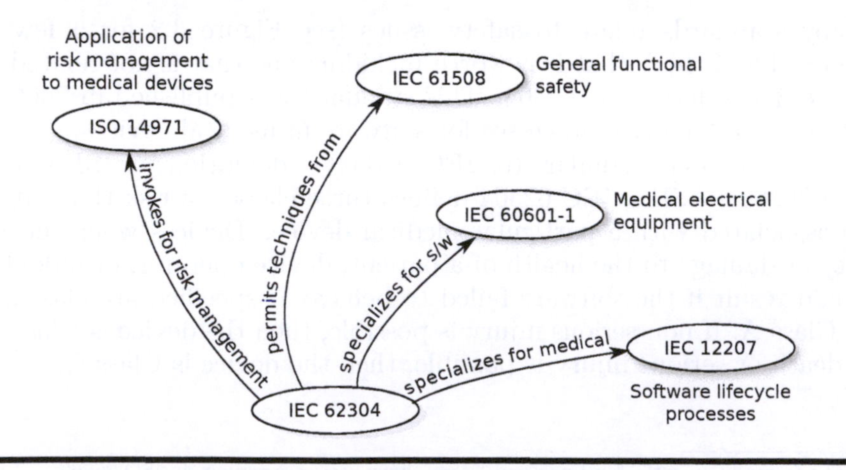

Figure 3.4 IEC 62304 and friends.

IEC 62304 focuses exclusively on processes and defines a number of these for the creation of software for a medical device:

Software development process.
This is broken down into eight subprocesses covering the steps from initial planning to software release.

Software maintenance process.
This covers effectively the same eight subprocesses.

Software configuration management process.
This includes the identification and correct handling of SOUP and the handling of requests for changes to the device.

Software problem resolution process.
This includes the obligation to inform users of the device of potentially dangerous bugs.

As illustrated in Figure 3.4, IEC 62304 specializes IEC 60601 for software and incorporates ISO 14971 ("Medical devices — Application of risk management to medical devices") to provide the hazard and risk identification and management process. This involves identifying hazards and risks, evaluating their severity, controlling them where necessary, and continuing to monitor the risks during the product's life. Generally, much of this information would be contained in a hazard and risk analysis — see page 56.

Process and the Standards

This book concentrates more on the technical than the procedural and process aspects of the standards. All of the standards we are considering do, however, define a process that should be followed when creating a software-based product.

In reference [8], Martyn Thomas makes a very shrewd observation about why good processes are important for a project:

> *What makes good processes essential is the confidence that they provide in their outputs: without well-defined testing processes you cannot interpret the results from testing; without strong version control you cannot have confidence that the product that was tested is the same one that was inspected, or integrated into the system. So strong, auditable processes are necessary if we are to have confidence in the product,* ***but the fact that a process***

> ***has been followed in developing a product does not tell you anything about the properties of the product itself.*** (emphasis mine)

Thus, strong processes need to be enforced during the development of a product, not because that will ensure that the product is of good quality, but because, without strong processes, we cannot trust any of the information we need to produce our safety case.

Fallacy 2 *A good process leads to good software.*

It is easier to measure adherence to a process ("the design document exists and was reviewed and issued before coding began...") than to measure quality of software. There is an implicit assumption in some of the standards, particularly IEC 62304, that the quality of the process is a measure of the quality of the end result. This is not so.

Summary

There are many standards relating to the development of software for deployment in safety-critical systems in avionics, railway systems, medical devices, industrial systems, robots, nuclear power stations, etc. These standards range from being highly prescriptive (do X and Y, and do not do Z), to being more goal-based (build a safe product and justify why you believe it to be safe).

This book focuses on IEC 61508, ISO 26262, and IEC 62304.

All three of these standards define different levels of criticality for the device being built and prescribe processes and techniques considered applicable to each level. In each case, the intention is to reduce the probability of device failure. This approach has been criticized by Nancy Leveson, who has counter-proposed a systems-theoretical rather than a reliability-theoretical approach to safety.

References

1. A. Rae, "Acceptable Residual Risk: Principles, Philosophy and Practicalities," in *2nd IET International Conference on System Safety*, (London), 2007.
2. J. McDermid and A. Rae, "Goal-Based Safety Standards: Promises and Pitfalls," in *2012 Safety Critical Systems Symposium*, SSS '12, (Bristol, UK), Safety-Critical Systems Club, 2012.
3. P. Reeve and M. Lloyd, "IEC 61508 and IEC 61511 assessments—some lessons learned," in *4th IET International Conference on Systems Safety 2009*, 2009.
4. M. Lloyd, "Trying to introduce IEC 61508: Successes and Failures," in *Warsaw Functional Safety Conference*, 25-26 February 2015.
5. N. G. Leveson, "System Safety Engineering: Back To The Future." Available at http://sunnyday.mit.edu/book2.pdf, 2008.
6. O. Nordland, "A Devil's Advocate on SIL 4," in *2012 Safety Critical Systems Symposium*, SSS '12, (Bristol, UK), Safety-Critical Systems Club, 2012.
7. AUTOSAR, "Requirements on Operating System," 2011. V3.0.0, R4.0 Rev 3, Final.
8. M. Thomas, "Engineering Judgement," in *SCS*, pp. 43–47, 2004.

Chapter 4

Representative Companies

You may know a man by the company he keeps.

To make the remaining chapters of this book less abstract, I introduce two imaginary companies in this chapter. We can then consider their tribulations while they design and develop their products for use in safety-critical systems, and follow them through their certification process.

Alpha Device Corporation

The **Alpha Device Corporation** (normally abbreviated to ADC) makes devices.

For the purposes of this book, it does not much matter to us whether these are medical devices being developed to meet IEC 62304/ISO 14971, automotive devices (ISO 26262), or industrial devices (IEC 61508), but for the sake of reducing abstraction, let us assume that they are automotive devices. Where there would be major differences between the development of medical and automotive devices these are pointed out as the story progresses.

In the terminology of the automobile industry, ADC is a "tier 1" supplier to an original equipment manufacturer (OEM), the company that actually makes the cars.

Of course, ADC does not create all the components in the device itself; both hardware and software components are bought from suppliers and integrated with ADC's own application code.

In particular, ADC intends to buy a major software component from **Beta Component Incorporated**, has informed Beta Component Incorporated of this, but has made the purchase contingent on Beta Component Incorporated achieving an ISO 26262 certificate stating that the component is suitable for use in "items" (the ISO 26262 term for a complete system to be fitted into a car) to be certified to automotive safety integrity level C (ASIL C).

ADC has gone one step further in its negotiations with Beta Component Incorporated and has named the certification body from which Beta Component Incorporated has to achieve its certification. As illustrated in Figure 3.1 on page 29, certificates from any certification bodies that have been accredited should be equivalent, but this is not always commercially so.

Beta Component Incorporated

Beta Component Incorporated (commonly known as BCI) makes software components to be used by companies such as ADC in mission- and safety-critical applications. Again, it does not really matter for our purposes what these components are; they might be databases, operating systems, communication stacks, toolchains, high-availability middleware or journaling file systems.

To make the discussion concrete, we will assume that BCI is to supply ADC with an operating system and is aware that the sale depends upon achieving an ISO 26262 certification for it.

As this operating system was not specifically developed for use in ADC's device, in the terms of ISO 26262 it is a "safety element out of context" (SEooC) — see page 41. Had ADC been a medical, rather than automotive device manufacturer, BCI's component would, in IEC 62304 terminology, be "software of unknown provenance" (SOUP).

In order to satisfy both the automotive and the medical markets, it's possible that BCI is working to achieve certification to both standards for its operating system.

Using a Certified Component

These descriptions of ADC and BCI are very simple, but contain one statement that might need expansion: "... made the purchase contingent on Beta Component Incorporated achieving an ISO 26262 certificate stating that the component is suitable for use in items to be certified to ASIL C."

Given that a certified component from BCI is likely to cost more than the equivalent uncertified component (which may be identical), why should ADC put this restriction on the purchase?

We shall assume that this is not, as is sometimes the case, simply an unthinking demand from someone within ADC's purchasing department who does not understand what the requirement means and has simply copied it from an earlier purchase order.

If this is a conscious decision, then ADC has performed a commercial calculation. By buying a certified component, ADC is not reducing its liability — the standards make it clear that the device manufacturer takes responsibility for all the components that make up the device, whether developed in-house or bought from a supplier.

This decision of whether to buy a certified or a standard component is explored more deeply in Chapter 6.

THE PROJECT II

Chapter 5

Foundational Analyses

Simple solutions seldom are. It takes a very unusual mind to undertake analysis of the obvious.

A. N. Whitehead

Analyses

I was tempted to entitle this chapter "Foundational *Documents*" because that's the way we often think about these activities — they result in documents. However, the goal is completion of the analyses, the documents recording those analyses being secondary.

The following list of analyses is not exhaustive, but those included may be considered as a foundation for the rest. They are each described in more detail below.

Hazard and risk analysis.

This identifies the hazards and risks associated with the use of the final product (for the distinction between a "hazard" and a "risk" and the meaning of "mitigation," see page 13).

This analysis is particularly crucial because it is from this that the project's safety requirements and data for the failure analysis are derived.

Safety case.

The phrase "safety case" uses the term "case" in the legal sense: "the case for the defense." It is a reasoned argument explaining why you believe the product you are developing is safe.

Failure analysis.
All systems eventually fail, and it is important to understand in what ways, and how often, your system is expected to fail.

In addition to the results of these analyses, there are two basic documents that form part of any development of a safety-critical application:

Safety plan.
This lays out the specific practices to be followed during the development of the product. For example, a company may have several coding standards, each suitable for a different type of coding; the safety plan specifies which is to be used for the project. It also defines the precise responsibilities of each project member.

Safety manual.
If you purchase any product, you are likely to find a safety manual attached — this is the sort of document that accompanies a purchased lawnmower and tells you not to try to use it to trim your beard. In addition to such liability coverage, in an embedded software product, it also defines the environment within which the product must be used and provides a list of proscriptions: e.g., "function `abc()` MUST* not be used unless at least 100 Mebibytes of free memory are available."

Interrelationships

These analyses are, of course, neither independent of each other nor carried out serially. Figure 5.1 illustrates some of the interrelationships:

- The hazard and risk analysis describes ways to reduce the identified risks, known as "mitigations," Each of these becomes a safety requirement. For example, if a particular risk can be reduced by installing duplicated power supplies, then this creates a safety requirement that two power supplies be fitted. Figure 19.2 on page 288 gives an example of deriving a safety requirement in this way.

* In requirements, words like SHALL, MUST, MUST NOT, MAY are often capitalized in accordance with RFC2119 (reference [1]). This avoids having to define the meanings of those words in each requirements document.

The hazard and risk analysis also identifies residual risks that are present even when the mitigations have been applied, and these become the basic elements of the failure analysis. For example, even with duplicated power supplies, there may be a residual risk that the power will be lost at a critical moment; this must be taken into account in the failure analysis.

- The hazard and risk analysis and the failure analysis become essential components of the safety case.
- The safety case defines the environment within which the product may be used and which features of the product may be used in different configurations. This information is captured in the safety manual that accompanies the product.

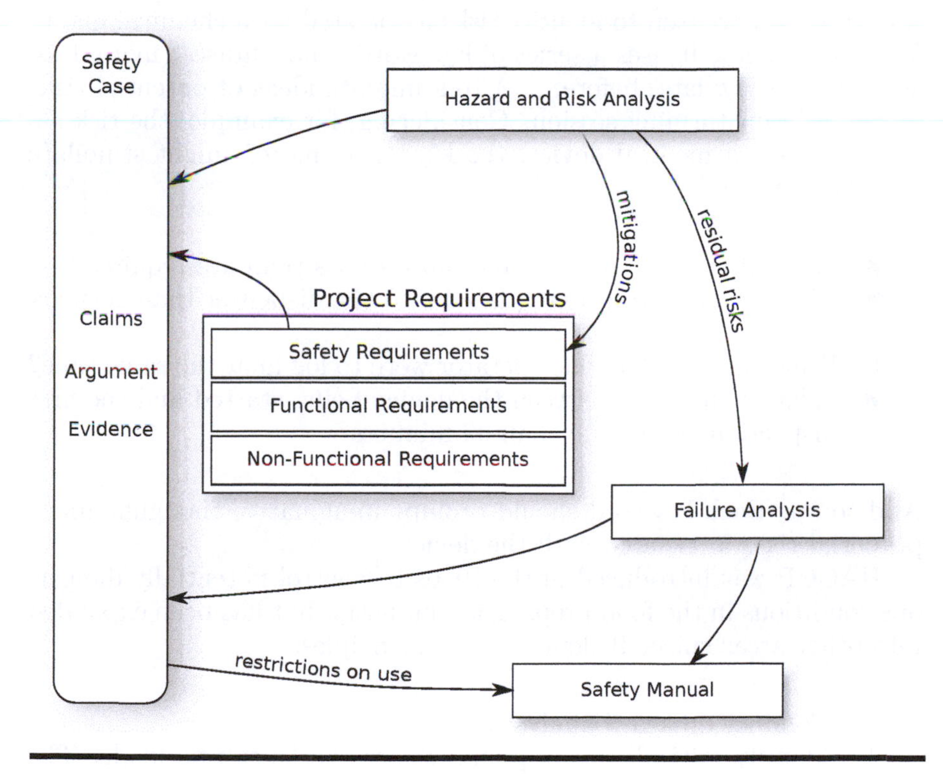

Figure 5.1 Interrelationship of the analyses.

Hazard and Risk Analysis

The idea of this analysis is to identify risks associated with the component or device, to determine mitigations to reduce those risks, preferably to zero, and then to determine what residual risks are left. ISO 14971, even though written for medical devices, provides a set of useful guidelines for identifying and mitigating hazards and risks and could be used to advantage even on projects not related to medical devices.

Identifying the Hazards and Risks

The first step in this process is to identify the hazards and the risks associated with them, and this is not easy.

There are structured ways of thinking about risks, for example, HAZOP ("hazard and operability study") and HACCP ("hazard analysis and critical control points").

The HAZOP procedure is defined in IEC 61882 (reference [2]) and was originally written to identify risks associated with chemical manufacturing plants. It uses a series of key words and phrases (more, less, as well as, early, late, before, ...) to stimulate ideas of potential risks during a brainstorming session. Considering, for example, the risk associated with a medical device, the keyword "more" might stimulate thoughts such as:

- What if the operator entered *more* digits than are required?
- What if *more* drug were placed into the dispenser than is specified?
- What if *more* than one operator were to log in at the same time?
- What if the time between the device being started and the first inspection were *more* than 30 minutes?

And so on. Each keyword should prompt imaginative thoughts about potential risks associated with the device.

HACCP was introduced in the 1960s to control potentially dangerous conditions in the food preparation industry, but has been expanded into other areas since. It defines seven principles:

1. Conduct a hazard analysis.
2. Identify critical control points where checks can be made. The idea is to avoid a single control at the end of the production chain where dangerous conditions may not be visible.
3. Set critical limits for each critical control point.

4. Define critical control point monitoring requirements to make sure that the control points are operating correctly.
5. Define corrective actions to be taken if a predefined critical level is exceeded.
6. Monitor the whole HACCP system to ensure that it is operating as expected.
7. Keep good records of the process.

ISO 14971 specifically mentions HAZOP and HACCP as suitable processes for identifying risks in medical devices.

Other useful sources of lists of potential hazards are Annex B of EN 50126-2, the railway standard, and ANSI/AAMI/IEC TIR80002-1:2009 (reference [3]), where Appendices B and C contain a list of risks and hazards to be considered when developing a medical device. TIR80002 is aimed at medical applications, but many of the risks listed are equally applicable to other areas. There are many of these, including "Function passes value in microlitres but driver expects value in millilitres," "Safe state never defined," and "Poor instructions for use cause the user to fail to calibrate the medical device correctly, which results in erroneous calibration constant."

Anecdote 4 *In 2014, I attended the presentation of reference [4] by Harold Thimbleby, at which he demonstrated the user interface on a number of portable medical devices that could reasonably be used by the same nurse during the course of a day.*

One point of the lecture was to demonstrate how something as superficially easy as typing in a numerical value (e.g., drug dosage) could differ so much from one device to another. For some devices, the number pad was arranged like a telephone (1, 2, 3 at the top); other devices had it arranged like a calculator (7, 8, 9 at the top). The DELETE or CLEAR buttons were also inconsistent in their behavior, and typing mistakes (e.g., entering a second decimal point in a number) were handled differently.

As Thimbleby said when he had demonstrated the ease of incorrectly entering a numeric value, not only is this a risk waiting to happen, it is a risk that has already happened and killed patients.

One type of risk that must not be neglected during the hazard and risk analysis is that arising from security hazards. In principle, once an attacker has exploited a security weakness to get access to a system, the behavior that can be caused is unpredictable and, in that sense, can be likened to random hardware failures or software Heisenbugs. Reference [5] by Simon Burton *et al.* gives some examples of functional security (as distinct from safety) requirements in the format of a safety case.

Mitigating Risks

Once a risk has been identified, what, if anything, needs to be done about it? Given that no system can be perfectly safe, is a particular risk acceptable? More generally, what is "safe enough"?

There are several ways of answering this last question, and the approach needs to be chosen and documented at the beginning of the project. The following approaches are often used.

ALARP.

This acronym stands for "as low as reasonably practicable" and the process is outlined in Annex C of IEC 61508-5 and described in detail in EN 50126-2 (Section 8.1 and Annex G). In essence, risks are assessed as belonging to one of three categories:

1. Risks that are clearly too great to be tolerated according to societal or other norms. Such risks must be reduced through mitigation.
2. Risks that are so small as to be acceptable within the society where the system is to be deployed. These risks, once identified and documented, must be monitored to ensure that they don't increase, but otherwise can be ignored.
3. Risks that lie between these two extremes. These risks must be reduced to the "lowest practicable level," balancing the benefits and costs of reducing the risk.

These regions are illustrated in Figure 5.2 and the boundaries of the regions for a particular project must be agreed before risk analysis begins. Annex G.2 of EN 50126-2 provides details of the actual regions used for the Copenhagen Metro development, but makes the point that every project will have different regions.

A risk, such as that labelled A, lies in the ALARP region and can be mitigated by reducing the level of harm it can cause (moving it to B), by reducing the probability of it occurring (to C), or both (to D).

Mitigating the risks in the ALARP region depends on the balance between cost of the risk and the cost of mitigating it. This often means making the unpleasant assessment of the cost of a human life or of environmental pollution. Thus, if a risk,

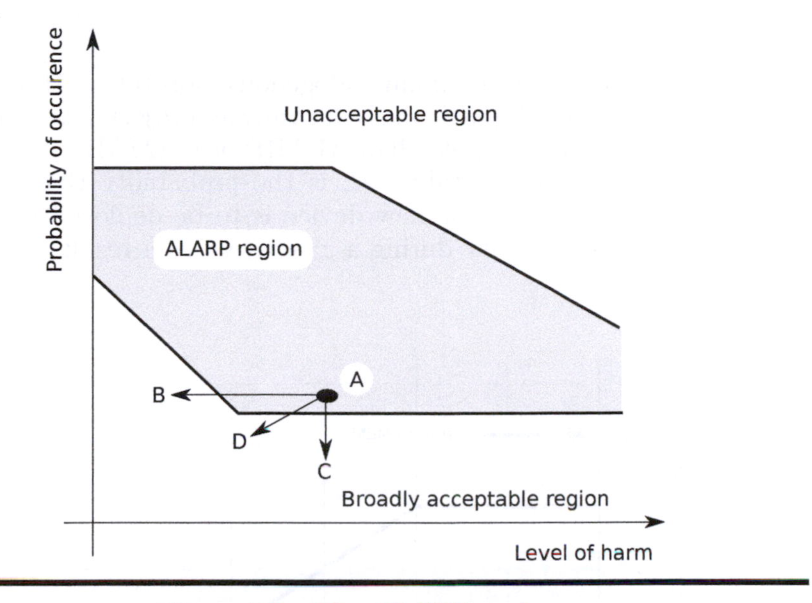

Figure 5.2 The ALARP region.

such as A in Figure 5.2, has been identified, and it has been determined that, if not mitigated, it is likely to lead to an accident once every 10 years, the cost to the company for the accident (say €10,000,000 in damages and loss of reputation) has to be weighed against the cost of performing the mitigation (say €1,500,000 per year).

Given those values, the mitigation would not be performed, as it is not cost-effective. For example, reference [6] quotes the UK Health and Safety Executive's figure of £1,000,000 (at 2001 prices) as a benchmark for the cost of one human fatality per year.

GAMAB.

An alternative to ALARP, used particularly in France, is *globalement au moins aussi bon (GAMAB).*

This method says that any new system must offer a level of risk, when measured by some acceptable indicator, no worse than the level offered by any equivalent existing system. Depending on the product, the indicator might be the number of casualties per year, the number of people injured per hour of operation, or some other suitable measure.

GAMAB is sometimes known as GAME *(globalement au moins équivalent),* although GAME does not incorporate the same idea of continuous improvement that is implicit in GAMAB.

MEM.

The principle of minimum endogenous mortality (MEM) is also sometimes used, particularly for railway projects, although it is less generally accepted than ALARP or GAMAB.

Endogenous mortality, R, is the probability that a person in the area where the new device is to be deployed will die or suffer serious injury during a given year as a result of "natural causes."

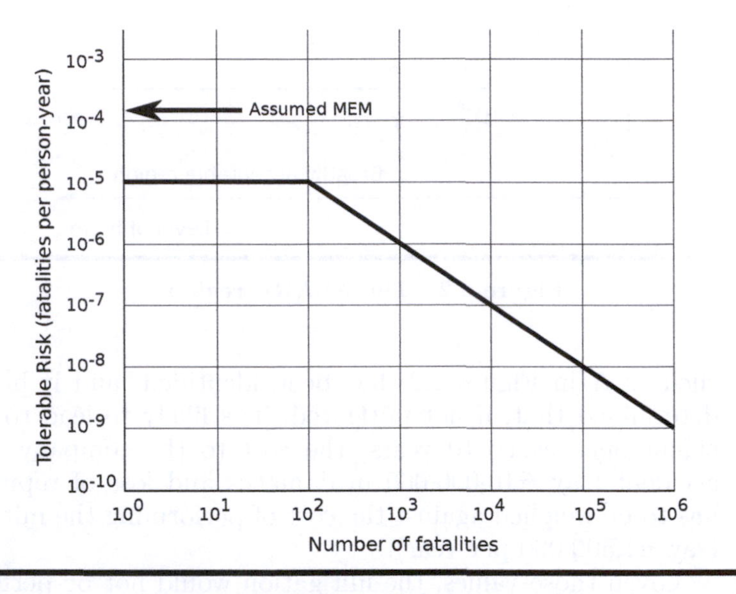

Figure 5.3 Differential risk aversion.

In developed countries R is at a minimum for the age group from 5 to 15 years and $R \approx 2 \times 10^4$ fatalities per person per year. MEM asserts that, for the target population, the new system should not "substantially affect this figure." This phrase is normally taken to mean, "should not increase it by more than 5%". This means that a new system deployed in a developed country should not increase the endogenous mortality by more than $0.05 \times 2 \times 10^{-4} = 10^{-5}$.

As illustrated in Figure 5.3, (taken from EN 50128), the 10% level has to be adjusted to take into account the number of fatalities likely to occur at one time because society is less tolerant of accidents involving many fatalities. In Figure 5.3, for example, it is assumed that the device is to be deployed in an area where the MEM is about 2×10^{-4} and that the permitted probability of fatality per year is about 10^{-5} for a small number of fatalities, decreasing to 10^{-9} for incidents involving one million people.

In these calculations, it is sometimes assumed that 100 light injuries or 10 serious injuries are equivalent to a fatality.

Assessing the Residual Risks

Even after a risk has been mitigated, there is probably some residual risk, and the mitigation process may need to be repeated.

In a few cases, mitigations can reduce the residual risk to zero, but normally some acceptable level of risk will remain.

Safety Case

The last section dealt with the hazard and risk analysis. Another essential analysis is concerned with the preparation of the safety case.

What is a Safety Case?

The safety case is a "a structured argument, supported by a body of evidence, that provides a compelling, comprehensive and valid case that a system is safe for a given application in a given operating environment." (from a UK Ministry of Defence definition, 2004).

For Whom Do We Create a Safety Case?

The obvious question is, "Whom are we trying to convince by our structured argument?" When I ask this question of engineers preparing a safety case, the knee-jerk answer often is "the auditor. We are preparing the case for the auditor so that we will get our certificate".

Frankly, I think that this is the wrong answer. Before the auditor sees the safety case, there are several other people that I believe it needs to convince.

The first are the authors of the safety case. If they don't really believe the argument, then it should not be put forward.

The second is the entire team working on the product development. If the members of the team don't believe in the safety of the product they are building, then the safety culture mentioned on page 4 doesn't exist within the company.

The third are the management and executives of the company building the product. As has been seen in several product recalls over safety concerns, the reputation of a company can be a very fragile asset. Shipping a product that damages or kills people can be both morally and financially expensive.

The fourth are the potential customers. The device under development will presumably be sold and the purchasers will need to be convinced that they are not exposing themselves to problems with the associated cost of recalls and bad publicity.

If these four groups are satisfied with the structure of the safety case argument and the evidence presented in it, then convincing an external auditor from the certification company is likely to be trivial.

I realize that this is a counsel of perfection unlikely to be fully achieved within any organization, but I believe that it is a target toward which we ought to strive.

Safety Case Problem

There is a deep problem associated with producing a safety case: confirmation bias. Confirmation bias, also called myside bias, is the tendency to search for or interpret information in a way that confirms one's existing beliefs or hypotheses. Nancy Leveson has pointed out (in reference [7]) that:

> *People will focus on and interpret evidence in a way that confirms the goal they have set for themselves. If the goal is to prove the system is safe, they will focus on the evidence that shows it is safe and create an argument for safety. If the goal is to show the system is unsafe, the evidence used and the interpretation of available evidence will be quite different. People also tend to interpret ambiguous evidence as supporting their existing position.*

This is confirmed by Sidney Dekker in reference [8], where he suggests that a better way of producing a safety case would be to ask the developers to prove that the system is *unsafe* — and hope that they fail.

That this attitude can prevail is made clear in the *Nimrod Review*, the official report, prepared by Sir Charles Haddon-Cave QC, into the crash of the Nimrod aircraft in 2006:*

> *. . . the Nimrod safety case was a lamentable job from start to finish. . . . Its production is a story of incompetence,*

* The full report is available from `https://www.gov.uk/government/publications/the-nimrod-review`.

> *complacency, and cynicism.*
>
> *The Nimrod Safety Case process was fatally undermined by a general malaise: a widespread assumption by those involved that the Nimrod was 'safe anyway' (because it had successfully flown for 30 years) and the task of drawing up the Safety Case became essentially a paperwork and 'tickbox' exercise.*

These two quotations remind us that, as we prepare the safety case, we should be conscious of our confirmation bias and ensure that the production of the safety case is an integral part of the development project, not an add-on.

Anecdote 5 *When teaching about confirmation bias in safety cases, I have several times used an experiment that I learned from reference [9] by Rolf Dobelli, although the original research was performed by Peter Wason in the 1960s. I write the numbers 2, 4, 6, 8, 10 on the whiteboard and tell the class that they are to deduce the rule behind the sequence. They can suggest numbers, and I will tell them whether the suggestion fits the rule or not.*

Inevitably, someone suggests 12, and I say that that conforms to the rule and add it to the sequence. Someone then suggests 14 and that also gets added. All of these suggestions are confirming the general impression that the rule is obvious. After a while, with perhaps 18 as the last number on the board, someone, possibly comically, may suggest 153. And I write that up. After some more suggestions, the class realizes that the rule is "the next number must be larger than the previous one" rather than "the next number is the previous number plus two."

The point is that the suggestions of 12, 14, 16, etc., which simply confirm what people already believe, have elicited much less information than suggestions, such as 153, that run counter to what is generally believed.

The same is true when preparing safety case arguments. Do we look for evidence to confirm our pre-existing belief that the system is safe, or do we specifically look for evidence that goes against our belief? The latter provides information that is more useful.

Safety Case Contents

The safety case consists of four parts (ISO 26262-10, Section 5.3 omits the first of these, but it is implicit in the rest of that standard):

Boundaries of the product.
What precisely is being included in the system for which the claims are being made? There are two different phases for which this needs to be defined, particularly if the system is just a component that is to be integrated into a larger system.

The first phase addresses the integration process: when we make claims are we, for example, including the driver that needs to be installed in the larger system to interface to our component? Is the debug interface covered by our claims?

The second phase deals with runtime: Are all the features provided by the component available for use, or are some prohibited by entries in the safety manual?

Claim.
What are we claiming about the device? This may include technical claims (e.g., "The device will respond within 50ms at least 99.999% of the time that it is invoked"), composite claims (e.g., "The device meets all of the requirements of ISO 26262 for a safety element out of context (SEooC) to be used in systems up to ASIL C"), or procedural claims (e.g., "The device has been developed in accordance with a development process that meets the requirements of IEC 62304 for devices of class C").

The claims must be explicit and verifiable.

Argument.
What convinces us that the product meets the claim? Although for a simple product, this argument might be written in a natural language, for any reasonably sized product, a semi-formal notation will be needed. The notations that are available include the *goal structuring notation* (GSN) and the *Bayesian belief network* (BBN). These notations are described in Appendices A on page 311 and B on page 315. In reference [10], Jörg Müller describe a further semi-formal notation for demonstrating that the processes involved in developing a product to a particular standard (EN 5012x in this case) have been met. I have not applied the notation myself and so cannot comment on its usefulness compared with the GSN and BBNs.

Whichever notation is used, it can be useful well before the audit to take the argument structure to the auditor who will finally certify the product and ask, "if I were to present the evidence to support this argument, would the argument convince you?" This is the opportunity for the auditor to say, "No, this part of the argument wouldn't convince me, you would have to provide additional justification here." Doing this early in the preparation of the safety case prevents wasted time gathering and arranging unconvincing or irrelevant evidence.

Evidence.

ISO 26262-10 points out that "An argument without supporting evidence is unfounded, and therefore unconvincing. Evidence without an argument is unexplained."

Once the structure of the argument is agreed, the evidence for each elementary claim ("leaf node" in the GSN or BBN graph) must be gathered and made accessible. A piece of evidence might be a document, a test result, a training record, a formal model, the proofs of theorems, or any other artifact that supports the argument.

If the argument is in HTML form, then it is convenient to include the links to the various pieces of evidence as soft links within the safety case: Click on the leaf node, and the artifact pops up.

There is one recursive problem with evidence as pointed out by Linling Sun and Tim Kelly in reference [11]. Does there have to be a subargument as to why that particular piece of evidence is adequate? Sometimes it is obvious. If it is claimed that a coding standard exists, then a link to the coding standard document suffices. But if the claim is that the development team is adequately trained, then does a link to the training records suffice? Is there not a need for an argument as to why those particular training records support the claim? Perhaps they refer to training that took place a long time ago or to training that is irrelevant to the current development.

This gap between the evidence and the claim that it is presented to support is known as the "assurance deficit." I believe that assessing the assurance deficit is one argument for using a quantitative notation, such as BBNs, rather than a qualitative notation such as GSN, for expressing the safety case argument. With a quantitative notation, the analyst can express the size of the deficit directly in the argument.

Each subclaim in the argument will probably have several pieces of evidence to support it. Consider the subclaim that "An adequate coding standard is employed during development." This would probably be broken down into individual subsubclaims:

A current coding standard exists.
The evidence here could be to the document itself* defining the coding rules.

The coding standard is of adequate quality.
The evidence here might be the record of the reviews held when the standard was first proposed, the record of reviews to which it has subsequently been exposed to keep it up to date, and the competence of the people who reviewed and approved the standard.

The coding standard is actually being used by the programmers.
Evidence for this may be harder to find. Perhaps an analysis of some modules taken at random could be made; perhaps developers could be interviewed, asking them about the coding standard; perhaps the output from a static analysis tool designed to check the coding standard rules could be made available; perhaps evidence that someone's code had been rejected for breaking a coding standard rule would be good evidence.

The example of the assessment of a coding standard is further developed in Figures B.4 (page 321) and B.5.

Retrospective Safety Case

As can be imagined, the easiest (and cheapest) way of producing a safety case is to make its creation an integral part of the development project — as artifacts are created, they are added to a database and take their place in the overall argument.

However, it is sometimes necessary to create a safety case for a product that already exists and for which no safety case was produced during the development. One advantage of retrospective development is that field history exists and it may be possible to base the safety case in part on proven-in-use (PIU) values: "The system has demonstrated t hours of field use with n failures having occurred; we therefore argue that the system meets the Y level of safety integrity".

An entire safety case can never be built out of PIU alone, but PIU data can support a powerful subargument as long as sufficient history can be demonstrated. Table D.1 in IEC 61508-7 provides a tabulation of the number of hours required to support a PIU argument for different safety integrity levels. As I explain in the description of this route (known as 2_S) on page 80, I believe that this table contains arithmetical

* The term "document" is used to include a paper document, a web page, a WIKI or whatever the company uses to control and publish processes to its engineers.

mistakes.

The question of proven-in-use is examined in the context of a non-certified component sourced from a third party on page 79.

Failure Analysis

The third foundation analysis, together with the hazard and risk analysis and the safety case, is the failure analysis.

All systems fail, and for safety-critical systems it is important to know how often and in what manner they fail. Essentially, this means considering the residual risks identified during the hazard and risk analysis.

In Figure 1.1 on page 8, failure analysis is shown as being carried out in two phases: a "quick-and-dirty" calculation to determine whether the proposed design could possibly meet the system's dependability requirements, the initial design cycle, and then a more sophisticated analysis, the detailed design cycle, to determine whether or not the system will meet its requirements. The former of these can be performed using a tool as crude as a Markov model (see Chapter 11), because the aim is not to determine whether the design *will* meet the requirements, rather the aim is to eliminate designs that *couldn't* meet the requirements.

The latter, more sophisticated, analysis can be carried out in one of three ways: bottom up, top down or event driven.

Bottom Up Approach

This takes each component of the system in turn and asks the question: "How often would this component fail and what would happen were it to fail?" This technique, known as *failure mode, effect and criticality analysis (FMECA),* is particularly popular when analyzing hardware failures. With hardware, it is easy to identify the components and their failure modes (e.g., Resistor R10 may fail by going open circuit or by going short circuit). The technique can also be applied to software, where it is sometimes known as *software FMECA (SWFMECA).*

FMECA has a number of limitations, including the combinatorial complexity that occurs when consequential failures are taken into account: R10 going short circuit will place additional stress on capacitor C23, thereby reducing its life and, if C23 shorts, then resistor R72 will heat up, ...

Another limitation of FMECA that is particularly relevant for software is the increasing difficulty of defining the failure modes of components. For a resistor, the modes are relatively clear (open or short circuit), but what are the failure modes of a system-on-a-chip (SoC) with 900,000,000 transistors incorporating a processor, memory and various coprocessors? How likely is it that the direct memory access (DMA) controller will fail while the other components continue to operate? How will that be detected? Software resembles the SoC much more than the resistor.

Within these limitations, the FMECA technique is quite straightforward:

1. List the components of the system.
2. Estimate the failure rate, λ_i, of each component. For electronic components various organizations issue failure rate estimates; for software components it is necessary to use more subtle techniques — see Chapter 13.
3. Identify the failure modes of each component and assign a probability ratio, α_i, to each. R10 can fail by going open circuit or going short circuit, but if the former is more likely than the latter, then $\alpha_1 = 0.7$ could be associated with the former and $\alpha_2 = 0.3$ with the latter.
4. Estimate the effect of failure of each component on the system's behavior. This is effectively a severity parameter, β, that defines the effect of failure on some scale (e.g., catastrophic, critical, marginal, negligible).
5. Determine how quickly the failure of each component will be detected. In the case of components whose failure causes the whole system to fail catastrophically, this may be relatively easy. Failure of components that only cause degradation or system failure under uncommon circumstances may remain hidden for a long time as time-bombs in the system.
6. For each component, calculate its contribution to the system failure: $C_r = t\beta\Sigma(\alpha_i\lambda_i)$, where t is the time for which the system must remain operational. Comparing the C_r value for the different components allows design improvement to be focused on the areas where there will be the most gain in dependability.
7. Estimate the time needed to repair the component once its failure has been detected. This will vary from component to component: those whose failure causes a system failure may be fixed as soon as possible after detection; those that cause a degradation might not be fixed until the next scheduled maintenance. Note that, except for cases of periodic replacement, repair cannot begin until detection has occurred — this is why detection is so important.

8. Build a mathematical model of the failure characteristics of the system (a combination of the components) and solve this model to determine the system's availability and its sensitivity to the various estimated parameters.

Fallacy 3 *It is often said (and repeated in many standards) that software does not fail randomly — all software failure is systematic and anyone analyzing software failure statistically falls into a state of sin. There are many academic papers explaining why this is a fallacy, but one simple program should be enough: see Figure 5.4.*

This program, when executed on any computer, particularly one with multiple cores, will crash with a floating point exception every few thousands of millions of times it is executed. The presence of other programs also executing on the same computer will change the probability of failure.

Top Down Approach

Whereas the bottom up approach says, "If this component were to fail, what dangerous condition might occur?", the top down approach starts with the dangerous conditions and asks, "What might cause this particular dangerous condition to occur?"

Using our representative companies from Chapter 4, the component company Beta Component Incorporated (BCI) might consider the potentially dangerous condition of the operating system running out of stack space. How might this occur? Perhaps if interrupt nesting were too deep. What is the likelihood that, given the protection (mitigation) provided, this could occur? Dangerous conditions on Alpha Device Corporation's (ADC) automotive device might include the simultaneous commands to accelerate and apply the brakes. This might occur if sensor A gave an over-reading and sensor B gave an under-reading within the same time window. What would cause sensor A to give that reading? If the speed were greater than 200 km/hr. And so on.

A top-down approach can be carried out using fault tree analysis (FTA). This is considered in depth in Chapter 12.

```
#include <stdio.h>
#include <pthread.h>

volatile int x = 0;

void * addOne(void *p) {
    int   y;
    int   i;

    for (i=0; i<100; i++) {
        y = x + 1;
        x = y;
        }
    return (0);
    }

int main() {
    pthread_t thread1, thread2;

    pthread_create (&thread1, NULL, addOne, NULL);
    pthread_create (&thread2, NULL, addOne, NULL);
    pthread_join(thread1, NULL);
    pthread_join(thread2, NULL);
    printf("%d\n", 18827/(x-5));
    return 0;
    }
```

Figure 5.4 A simple program.

Anecdote 6 *The example above of a car simultaneously accelerating and braking is not completely theoretical. I once had a student working with me who needed an example of a system that she could formally prove to be correct, using the* Rodin *tool; see page 225.*

In a technical journal, she found an article by a leading car manufacturer describing its collision detection and cruise control systems. We decided that it would be a good exercise for her to model this with the intent of proving a number of characteristics of the design, including the fact that there was no possible way that the system could simultaneously command braking and acceleration. Rodin was unable to prove this invariant, and the reason became clear when we analyzed the design — the design was flawed and, under unusual conditions, this dangerous condition could occur.

Event-Centric Approach

Both the bottom-up and top-down approaches described above deal with components of the system, albeit from different directions. An event tree analysis (ETA), in contrast, starts with events. Figure 5.5 illustrates a very small part of an event tree that might be associated with a medical device.

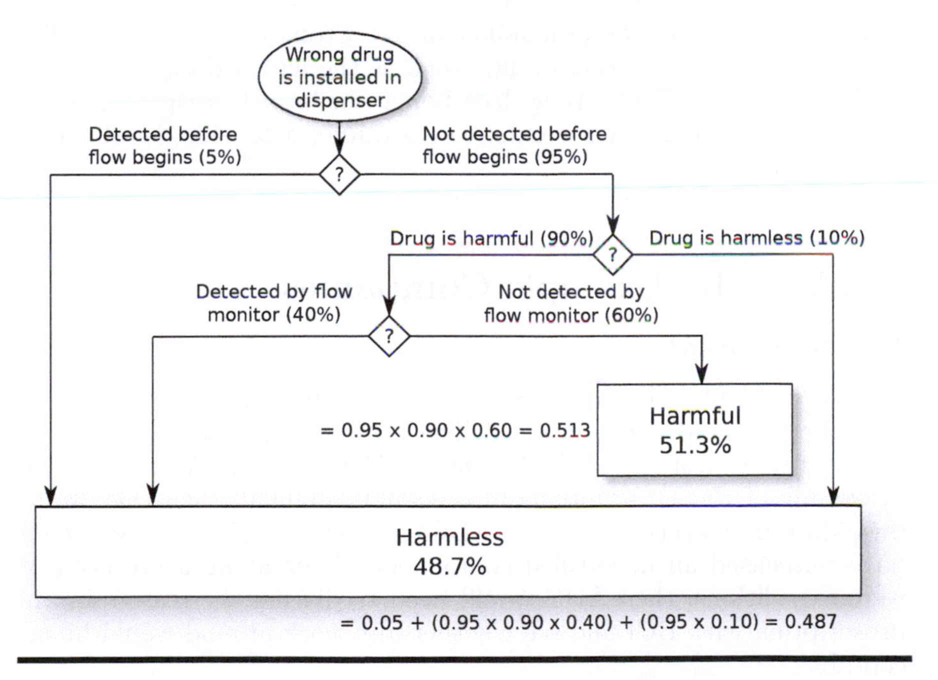

Figure 5.5 Event tree example.

The event being considered in this example is the wrong drug being put into the dispenser. This potentially hazardous situation might be detected by the operator before flow begins (assessed to be 5% probable), and, even if the flow starts, the drug might be harmless to the patient (10%). Even if the flow starts and the drug is harmful, the flow monitor might detect that the wrong drug is being dispensed, based perhaps on its viscosity (40%).

When probabilities have been measured and added to the event tree, it is easy to calculate the probability of a dangerous outcome: In this case 51.3% of the times that the wrong drug is put into the dispenser.

The effectiveness of this technique depends on the system being sufficiently well-understood for a comprehensive selection of potentially dangerous events to be identified. The HAZOP procedure defined in IEC 61882 and described on page 56 has been used as a semi-structured way of identifying these events. If HAZOP has been used to perform the hazard and risk analysis, then the residual risks from that can be used to identify potentially dangerous events.

One disadvantage of ETA is its implicit assumption that the identified events are independent of each other. In fact, this is unlikely to be true: Perhaps the top-level event in Figure 5.5 is dependent on another event such as: "Nurse is required to work second consecutive shift." That could increase the probability of the wrong drug being installed in the dispenser, and other events, such as the wrong dosage being entered. The event of the wrong drug being put into the dispenser might also originate from other events; for example, a labelling problem in the dispensary.

Analyses by Example Companies

Safety Manuals

For the example companies introduced in Chapter 4, Beta Component Incorporated (BCI) would deliver a safety manual, together with the operating system, to Alpha Device Corporation (ADC). If ADC broke any of the proscriptions in the safety manual, then that would invalidate BCI's certification for ADC's device: ADC would effectively have purchased an uncertified component. If breaking a proscription were expeditious, then ADC would need to discuss the reason for the proscription with BCI and see whether any work-around would be acceptable.

Also ADC would publish a safety manual with its finished device providing its customers with limits on its use: e.g., "this device must not be used above 5,000 metres above sea level."

Hazards and Risks

BCI might have identified the following risk:

> **BCI Risk 0734:** If a memory error corrupts the configuration data holding the scheduling policy, the operating system could silently schedule threads incorrectly, causing the application to behave unpredictably.

ADC's risks would be at a higher, device, level and might include:

> **ADC Risk 1007:** If the driver selects mode 3 on the device while braking, the delay in recognizing the mode change could allow engine power to be incorrectly increased.

Safety Requirements

We will assume that each of the risks identified above will need mitigation. This leads to BCI's first safety requirement:

> **BCI Safety Requirement 0734-A:** All configuration data SHALL be stored in such a way that memory corruption can be detected.

Facing this requirement, BCI's design team considers various ways of implementing it and decides that putting a checksum on the configuration data and checking it every time the data are used would create an intolerable performance overhead (the balance between safety and usability — see page 91). After more brainstorming, the team decides to store each value, not as a single bit, but as a word containing distinctive bit patterns, but there is still a concern that a good, optimizing compiler might notice that the values are only being used as Booleans and optimize them back to single bits. This creates two further safety requirements:

> **BCI Safety Requirement 0734-B:** The values of each configuration datum SHALL be stored as distinctive multibit values such that no single or double bit corruption would lead to another valid value.

> **BCI Safety Requirement 0734-C:** The means of storing multibit values in accordance with requirement 0734-B SHALL be such that the compiler does not reduce them to a single bit, irrespective of the optimization level used.

Note the numbering scheme — it is necessary to demonstrate that each identified risk has been mitigated (or consciously not mitigated), and using a common numbering scheme simplifies the tracing.

Residual Risks

In the case of BCI's configuration data, the accepted mitigation may not be impervious to triple-bit errors.

These residual risks will form part of the failure analysis: see page 67. From BCI's point of view, statistical information about the frequency and types of memory errors in DRAMs could be used to estimate the probability per hour of use of an undetected corruption, and this figure could be used in the failure model.

Safety Case

Both ADC and BCI will have to prepare the analyses for their safety cases. Whether BCI delivers its safety case to ADC as part of its product delivery is a purely commercial decision — ADC may require it as part of the contract.

Summary

While designing and implementing a software-based device designed for a safety-critical application, there are numerous analyses that need to be carried out. The central of these is the preparation of the safety case, the presentation of the argument as to why the device is considered to be safe. This will contain a clear statement of the claims made about the device, an argument justifying its use in safety-critical applications, and the evidence backing the argument.

The safety case will incorporate the results of other analyses, including the hazard and risk analysis and the failure analysis.

References

1. S. Bradner, "RFC2119: Key words for use in RFCs to Indicate Requirement Levels," 1997.
2. IEC, "IEC61882: Hazard and operability studies (HAZOP studies) — Application guide," 2001.
3. ANSI/AAMI/IEC, "Medical device software — Part 1: Guidance on the application of ISO 14971 to medical device software," tech. rep., Associa-

tion for the Advancement of Medical Instrumentation, Arlington, VA, USA, 2009.

4. H. Thimbleby, "Safety versus Security in Healthcare IT," in *Addressing Systems Safety Challenges, Proceedings of the 22nd Safety-Critical Systems Symposium* (C. Dale and T. Anderson, eds.), pp. 133–146, 2014.
5. S. Burton, J. Likkei, P. Vembar, and M. Wolf, "Automotive Functional Safety = Safety + Security," in *Proceedings of the First International Conference on Security of Internet of Things*, SecurIT '12, (New York, NY, USA), pp. 150–159, ACM, 2012.
6. Rail Safety and Standards Board, "Engineering Safety Management (The Yellow Book)," 2007. Available from `www.yellowbook-rail.org.uk`.
7. N. Leveson, "White Paper on the Use of Safety Cases in Certification and Regulation," 2011.
8. S. Dekker, *The Field Guide to Understanding Human Error*. Ashgate, Aldershot, Hants., rev. ed. ed., 2006.
9. R. Dobelli, *Die Kunst des klaren Denkens*. München: Deutscher Taschenbuch Verlag GmbH, 2014.
10. J. R. Müller, J. Drewes, J. May, and C. Trog, "The Formal Representation of the Safety Case Processes described in the EN 5012x norms," in *International Railway Safety Conference (IRSC)*, (Baastad, Sweden), 2009.
11. L. Sun and T. Kelly, "Elaborating the Concept of Evidence in Safety Cases," in *Proceedings of the 21st Safety-Critical Systems Symposium*, pp. 111–127, Safety-Critical Systems Club, 2013.

Chapter 6

Certified and Uncertified Components

Je vis de bonne soupe, et non de beau langage.

Molière

A system is built from components. This chapter considers some of the aspects of incorporating external components into a system that will be subject to certification. I use the terms "integrator" and "component supplier" to refer to the team building the system, and the source of the components, respectively.

Note that the component supplier may be another project within the integrator's company, an external commercial vendor, or an open-source project.

SOUP by Any Other Name

IEC 62304 has the best terminology for these types of components: SOUP, or *software of unknown provenance,* although sometimes "uncertain" is substituted for "unknown" and "pedigree" for "provenance," thus providing four expansions of the single acronym.

IEC 61508's equivalent of SOUP is "pre-existing software element," whereas in ISO 26262, it is termed a "safety element out of context" (SEooC) as described on page 41.

IEC 62304 defines SOUP as being either of two types:

1. A software item that is already developed and generally available and that has not been developed for the purpose of being incorporated into the medical device.
2. Software previously developed for which adequate records of the development processes are not available.

The first of these two definitions effectively means that any pre-existing component brought in for use in a product must be treated as SOUP, irrespective of how it was developed and whether or not it has been certified.

Certified or Uncertified SOUP

For many of the components that are included into a certified system, there is no choice between using a certified or uncertified version because the component supplier cannot supply a certified version. This is true of many very useful and popular open-source products, including Linux, Apache, and MySQL.

Where a component supplier has both certified and uncertified versions of a component, the question is whether the integrator should buy the commercial-grade product that does not carry a certificate, or the certified version that is normally more expensive.

There are two questions that the integrator needs to answer when considering this type of purchase. The first is technical: "How much reliance does the design place on the component to maintain functional safety, and can the design be changed so that less reliance is placed on the component without unacceptably hitting product cost, verification cost, or delivery date?"

The second question is one purely for project management: "If a particular component is essential for maintaining functional safety, will it cost more in money, project risk, and project duration to prepare the certification evidence in-house or to pay the premium for a pre-certified component?"

For automotive systems, paragraph 5.4.10 of ISO 26262-9 explicitly permits, under certain stringent conditions, the decomposition of a system into subsystems of different ASILs. For example an ASIL C system could be decomposed into an ASIL C part and a QM(C) component (i.e., a "quality management" component for which it is only necessary to demonstrate that it was developed in accordance with an adequate quality management system). It may be easier to demonstrate this level of quality management for a third-party component, than to demonstrate that it meets all the requirements of ISO 26262 by buying a certified component.

Using Non-Certified Components

The standards make demands on bought-in components. Depending on the particular standard and safety level, many pieces of evidence will be required before the integrated device can be certified.

Of the standards we are considering in this book, IEC 61508 lays out the clearest track. In paragraph 7.4.2.12 of part 3, three possible routes are provided for integrators incorporating pre-existing software elements:

1. Route 1_S: development compliant with IEC 61508.
2. Route 2_S: proven-in-use (PIU).
3. Route 3_S: assessment of a noncompliant development.

The strange nomenclature for these routes is explained in IEC 61508-2: the S indicates software, to distinguish these routes from the hardware equivalents, 1_H, etc.

All three of these routes are specific to IEC 61508. Both IEC 62304 and ISO 26262 have their own ways of incorporating SOUP and SEooCs, respectively. I include comments on these as appropriate in the descriptions of the IEC 61508 routes below.

Route 1_S

Route 1_S is only applicable if the component was originally developed in accordance with IEC 61508 and all the certification artifacts (documents, analyses, safety manual, etc.) exist. The component may or may not actually have been certified and, if not, this route simply gathers all the artifacts together and effectively certifies the component at the same time as the system into which it is incorporated. In most cases, this is a completely impractical route for non-certified components; very few non-certified components were originally developed in such a way that all the artifacts were created, stored for later use, and kept up to date.

Route 2_S

Route 2_S requires that the integrator demonstrate that the component has been used for a certain number of hours in a role equivalent to that in which it will be used in the integrated product, with available records of failures and usage. Section 7.4.10 of IEC 61508-2 provides the outline of the case that has to be presented to demonstrate route 2_S and starts with the warning:

> *An element shall only be regarded as proven in use when it has a clearly restricted and specified functionality and when there is adequate documentary evidence to demonstrate that the likelihood of any dangerous systematic faults is low enough that the required safety integrity levels of the safety functions that use the element is achieved. Evidence shall be based on analysis of operational experience of a specific configuration of the element together with suitability analysis and testing.*

It might be difficult to demonstrate that a sophisticated, multipurpose component, such as Linux, has "clearly restricted and specified functionality."

The meat of the requirements for demonstrating that a component has been proven through use is contained in annex D of IEC 61508-7. The approved formula for calculating the required length of fault-free operation for a continuously operating component is

$$t = \frac{-\ln(\alpha)}{\lambda} \tag{6.1}$$

where t is the required time, $1 - \alpha$ is the required level of confidence, and λ is the failure rate per hour* that is being claimed. For a safety integrity level 3 (SIL 3) system, for example, $\lambda < 10^{-7}$, and so, for a 95% confidence that the system is proven through use, at least

$$t = \frac{-\ln(\alpha)}{\lambda} = \frac{-\ln(0.05)}{10^{-7}} = 29,957,323 \text{ hours} \tag{6.2}$$

of in-use operation are required.

Note that this result differs by a factor of 10 from the worked example in Table D.1 of IEC 61508-7, which I believe to be a misprint

* Actually, the calculation in IEC 61508 assumes that the failure rate per hour is the same as the probability of failure per hour — this is a valid approximation for low failure rates.

(it appears that the authors of Table D.1 used $\lambda = 10^{-8}$ rather than $\lambda = 10^{-7}$ as the bound on SIL 3). 29,957,323 hours is about 3417 years and is not an unreasonable number of hours for an embedded device that has been shipped in quantity for some years. Of course, making such an argument means not only having the records to demonstrate the usage hours, but also details of all field failures.

For systems that do not have to operate continuously but only on demand, the probability of failure on demand (p) is related to the confidence level $(1 - \alpha)$ and the number of field operations (n) by the relationship

$$n = \frac{-\ln(\alpha)}{p} \tag{6.3}$$

As an example, to provide evidence at the 99% confidence level ($\alpha = 0.01$) that a system meets the demands of SIL 3 for low demand mode ($p < 10^{-3}$) requires evidence of 4605 correct operations.* For a low-demand mode system, where operation is assumed to occur less than once per year, such evidence may be hard to find: requiring thousands of years of field experience.

Route 3_S

Route 3_S is described in detail in paragraph 7.4.2.13 of IEC 61508-3 and effectively means that the integrator, or the component supplier under the integrator's direction, retrospectively performs the design and verification required to satisfy IEC 61508 — defining the safety requirements, justifying the use of the component, documenting the component's design, recording the verification and validation performed on the component, performing a hazard and risk analysis, performing a failure analysis and preparing a safety manual.

IEC 62304 additionally requires (section 5.1.7) that a hazard and risk analysis be performed on the component. Even for an ISO 26262 or IEC 61508 development, it would probably be wise to carry out a hazard and risk analysis.

In summary, route 3_S can entail an enormous amount of work. Even preparing a safety manual for a component as complex as Linux or MySQL** could take significant time. In this regard, IEC 61508 states:

---

* Again, there seems to be an arithmetical slip in Table D.1 of IEC 61508-7.

** I use Linux and MySQL as examples in this section because they are well-known and well-proven pieces of technology that could well form part of a system design. The same argument applies to many other pieces of open-source or commercial off-the-shelf software.

> *The safety manual may be derived from the element supplier's own documentation and records of the element supplier's development process, or may be created or supplemented by additional qualification activities undertaken by the developer of the safety related system or by third parties. In some cases, reverse engineering*[***] *may be required to create specification or design documentation adequate to meet the requirements of this clause, subject to the prevailing legal conditions (e.g. copyright or intellectual property rights).*

As well as retrospectively creating a safety manual, there are other significant tasks to be considered. The integrator will need to prepare evidence to show the following.

The assumptions the component supplier made about the use of the component meet the actual use to which it will be put.
Perhaps the component supplier only intended the component to be used at interrupt rates lower than 1000 per second, but in the integrated device, higher interrupt rates might occur.

ISO 26262-10 contains a diagram to illustrate the interface between the component (SEooC) supplier and the integrator. I have adapted that diagram in Figure 6.1 which shows that it is the responsibility of the SEooC developer to define the assumptions made about the use to which the component will be put and the responsibility of the item developer to ensure that those assumptions are valid. Those assumptions would normally be delivered in the SEooC's safety manual.

The failure modes and failure rates of the component have been analyzed.
The component will appear in the failure analysis of the integrated device, and information about its failure modes and failure rate will be needed.

Impact analyses are being performed on bugs found in the component.
The question asked during the impact analysis is, "Could any of these bugs affect the safe operation of the device?" This is an activity that will continue for the life of the device and could be difficult with an open-source product where the bug lists are freely accessible, although sometimes only interpretable

[***] That is generating the design document from the implementation rather than the implementation from the design.

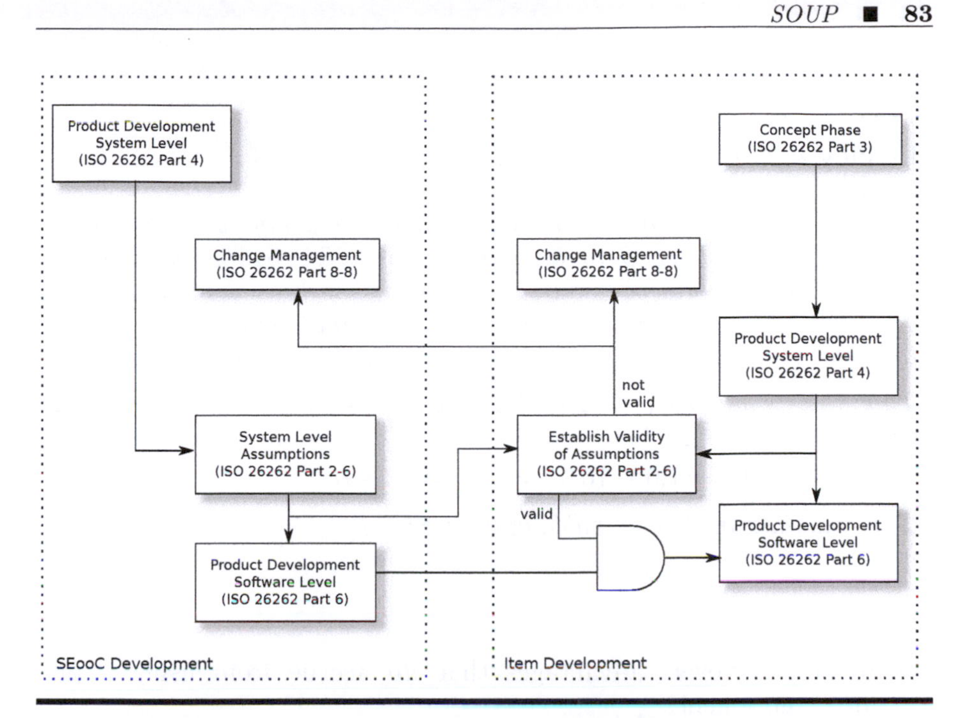

Figure 6.1 SEooC and item development (ISO 26262-10).

by an expert, and potentially impossible for a closed-source, commercial product.

This level of scrutiny of bug reports is also demanded in Section 7.1.3 of IEC 62304.

This is what deters some integrators from using open-source software in safety-critical systems — producing a failure analysis for a product such as Linux could be prohibitively expensive and would have to rely on expertise that the integrator might not have. Similarly, tracking the Linux bug reports, checking every one for potential impact on the product throughout the product's life, could be very time-consuming.

Using a Certified Component

If the supplier has had the component certified, then route 1_S becomes the most practical. Definite advantages come from using a certified component, but there is one trap to avoid. It is sometimes believed that "by buying certified components, we can be sure of certification of the final product." See Fallacy 4 for a quick refutation of that idea.

Fallacy 4 *SIL 3 + SIL 3 = SIL 3*

It is tempting for companies to believe that if they buy two components, each certified to SIL 3 in accordance with IEC 61508, so that the probability of failure per hour of operation has been demonstrated to be $< 10^{-7}$*, and combine these into a system that relies on them both, then the system is also suitable for SIL 3 certification.*

That this is not so is made clear by means of a simple arithmetic calculation:

$$1.0 - (1.0 - 10^{-7})(1.0 - 10^{-7}) = 2 \times 10^{-7} > 10^{-7}$$

The combination is certifiable only at SIL 2.

There are, however, advantages that *do* accrue from using a pre-certified component.

First, if the component's certificate is accepted by the integrator's certification authority, then routes 2_S and 3_S can be avoided. In principle, if the company that issued the component's certificate is accredited (see Figure 3.1 on page 29), then it would be expected that any other accredited certification authority would accept it. I have not always found this to be the case; sometimes one certification company is suspicious of a component certificate issued by another. It is not clear whether this is commercially driven or due to a genuine concern for the quality of the other company's certificate.

Secondly, the conditions of the certification will place an obligation on the component supplier to keep the integrator informed of any bugs that are discovered in the component that could potentially affect safe operation. For a non-certified component, this information may not be readily available and it is not clear how a product that relies on a component can satisfy a safety function without this information.

One advantage does *not* come with the purchase of a certified component — using a certified component does not reduce the integrator's moral or legal responsibilities. IEC 62304 spells it out most clearly:

> *Therefore, when the medical device system architecture includes an acquired component (this could be a purchased component or a component of unknown provenance), such as a printer/plotter that includes SOUP, the acquired component becomes the responsibility of the*

manufacturer and must be included in the risk management of the medical device.

Note that this not only includes software components such as MySQL, which are integrated directly into the product, but also embedded code inside devices sourced from a third party. When preparing such analysis, we need to be careful about "accidental systems" of the type described on page 19. In that case, it may be very difficult to identify and gain access to all of the code.

Aligning Release Cycles

Whether a certified or uncertified component is used, the question of integrating release cycles needs to be addressed. Each component may have a different release cycle, and these release cycles need to be coordinated with the release cycle of the integrated product. Section 6.1 of IEC 62304 also reminds us that we need to accommodate the possibility of the third-party component becoming unsupported; if the organization producing it disappears or stops supporting the component, how will that circumstance be handled by the quality management system (QMS) of the integrator?

Example Companies

For the fictitious Alpha Device Corporation introduced in Chapter 4, the decision regarding buying the certified or uncertified version of Beta Component Incorporated's operating system seems to be clear — ADC intends to buy the certified version. This decision was based on the technical observation that the operating system is crucial for maintaining functional safety in the device, and on the commercial consideration of the risk to the end date of the development if the uncertified version were used.

This is not the end of the story. ADC's quality management system defines the way in which third-party components may be incorporated into a product. These include performing an audit of the supplier's development processes and, to mitigate the risk of BCI stopping support for the component, access to the source code and build system for the component.

The contractual agreement between ADC and BCI, therefore, includes an option for ADC to carry out an on-site audit of BCI's development procedures at least once a year. This may be undertaken by staff from ADC's quality management group or by an external auditor selected by ADC. This requirement also adds cost to the operating system because BCI will have to dedicate staff to the audit each year.

The question of access to the source code of the operating system is more contentious because this is part of BCI's intellectual property. After much negotiation, an agreement is reached, whereby the source code and the infrastructure to build it will be put into escrow — a neutral place where ADC can get access under predefined conditions, such as BCI going out of business or stopping support for the component. This agreement also demands that, as BCI makes bug fixes to the component, the version in escrow will be kept up to date.

DESIGN PATTERNS

Chapter 7

Architectural Balancing

> *The bottom line for mathematicians is that the architecture has to be right. In all the mathematics that I did, the essential point was to find the right architecture. It's like building a bridge. Once the main lines of the structure are right, then the details miraculously fit. The problem is the overall design.*
>
> Freeman Dyson

The chapters in this part of the book describe a number of architectural and design patterns. The main patterns described are:

- those useful for error detection, including anomaly detection. These are covered in Chapter 8.
- those for handling errors should they occur; see Chapter 9.
- those for replication and diversification, including the "safety bag" or diverse monitor pattern; see Chapter 10.

Selecting from amongst these patterns means acknowledging that many architectural balances have to be achieved. Many characteristics of a system pull against each other — the architect satisfies one at the expense of another — and these trade-offs need to be consciously taken into account and the justifications recorded.

Reference [1] by the present author and Akramul Azim provides some examples of these tensions and points out the scarcity of tools that the analyst has available to help resolve them.

Availability/Reliability Balance

The distinction between availability and reliability is described on page 14. Effectively, "availability" means that the server responds in a timely manner, whereas "reliability" means that the server's response is correct.

It is possible to imagine systems where safety is preserved by high reliability and others where safety demands high availability. Systems that work in offline mode lie in the first category. For an automated medical device that analyses patients' blood and urine samples in a laboratory, away from the actual patient, reliability would be of higher importance than availability. If the device failed completely from time to time this would be frustrating and put the device at a commercial disadvantage, but it wouldn't endanger a patient. If it were unreliable and gave incorrect results, particularly false negatives,* this could endanger the patient.

For many systems that provide functional safety, occasional unreliability can be tolerated, while unavailability is dangerous. This is particularly true of systems where continuous operation is required and where self-correcting algorithms have been used to reduce the chances of a dangerous situation occurring. The Kalman filter — see page 106 — is an example of such an algorithm. If occasional values are unreliable, then they may work their way out of the algorithm over time without any dangerous condition having occurred. In this case, unreliability may be tolerated, whereas an unavailability might be unacceptable.

Anecdote 7 *When I start working with a company on the design of a piece of safety-critical equipment, the first questions I tend to ask are, "How often may it fail in a dangerous manner?" and "Which is the more important: the system's availability or its reliability?"*

The initial answers are often "it must never fail" and "it must be available 100% of the time and must be 100% reliable."

These are not helpful answers.

That a balance is required between availability and reliability is particularly clear when replication or diversification is used in the architecture (see Chapter 10). Presenting the same information to two subsystems

* Indicating that the sample showed no sign of disease when actually it did.

can improve the reliability of the result if the two outputs are compared. This two-out-of-two (2oo2) design, however, reduces system availability because the failure of either subsystem renders the complete system inoperative. If rather than comparing both outputs, the first output is accepted, then the availability of the system increases, but the reliability decreases.

Usefulness/Safety Balance

I have deliberately used the term "usefulness" in this section to avoid overloading the term "availability," which would perhaps have been a more conventional word. Consider a device that automatically stops a train when some event occurs. If, due to a fault in the device, it stops the train unnecessarily, the device can still be considered "available" because it has actively kept the train safe. The unnecessarily stopped train is, however, not useful.

It is trivially easy to build a very safe system. A train that never moves is perfectly safe, an aircraft that doesn't take off can be safe, traffic lights that are permanently red are safe, at least until the drivers lose patience and start to ignore them.

The balance that has to be achieved is that between usefulness and safety, and this is where a highly reliable system (e.g., 2oo2) can become unacceptable. A device moving to its design safe state generally becomes useless, but worse, it puts stress onto the larger system (its environment) into which it has been incorporated. This is particularly dangerous when the larger system is an "accidental system" of the type described on page 19. In that case, the effect of the move to the design safe state is unpredictable.

Even if the device's move to its design safe state is accommodated within the larger system, the move presumably only happens occasionally and so may not be well-proven. In particular, for systems where a human operator is involved, the device moving to its design safe state frequently can cause a human operator to start working around the safety measures: jamming interlocks open, etc.

For these reasons, the balance between usability and safety needs to be considered carefully; there are actually four states of the system as listed in Table 7.1.

State A is, one hopes, the normal operating mode of the device: There is no dangerous condition and the system is operating. State B occurs when the system moves to its design safe state because it believes that there is a dangerous condition when in fact there isn't. In state B, the system is safe, but useless. State C is the most worrying of

Table 7.1 System states.

State		Safe?	Useful?	Stress on its Environment?
A	Operating, no dangerous condition	Y	Y	N
B	Safe state, no dangerous condition	Y	N	Y
C	Operating, dangerous condition	N	N	Y
D	Safe state, dangerous condition	Y	N	Y

the states; this state occurs when the system fails to move to its design safe state when a dangerous condition occurs. State D occurs when the dangerous condition is detected and the device correctly moves to its design safe state.

Notice that the system is useful in only one of the four states, whereas it is safe in three of them. The highly reliable 2oo2 system described above might move to state B too often to be acceptable to the customer if its functionality is too hair-triggered, moving to its design safe state whenever there is a momentary disagreement between the two subsystems. Following the usability analysis, if such a possibility exists, it may be possible to permit one of the subsystems to restart and resynchronize following a disagreement between the two processors. If this can be done quickly, it may not undermine the safety of the system. In, say, 100 ms, a high-speed train moves 10 metres and a drug-dispensing device releases 3 μl of drug. If one of the subsystems can repeat the calculation during that time, it might be better from the safety point of view to allow it to do so, because systems that move to their design safe state too readily can, in themselves, be a hazard. This is something that would need to be addressed in the hazard and risk analysis and in the failure analysis.

Security/Performance/Safety Balance

During the last few years, the topic of security has become intertwined with the topic of safety. A decade ago, most systems that provided functional safety were kept secure by physical isolation — they were physically locked into the cab of a railway train, or in an embedded electronic control unit (ECU) in a car or a hospital's operating room.

This innocent state of affairs is changing rapidly, and today almost every device provides many entry points for a malicious attacker. USB ports, operator keyboards, Wi-Fi, Bluetooth, and, in cars, wireless tyre pressure connections provide points of direct access for attackers; reliance on the integrity of external signals, such as the global positioning system (GPS) and the global system for mobile communications (GSM), provides attack surfaces for jamming or spoofing.

Newspapers and research papers are already reporting systems within cars being taken over by roadside hackers. See, for example, reference [2] by Karl Koscher *et al.*

As vehicle-to-vehicle (V2V) and vehicle-to-infrastructure (V2I) communication becomes common for collision avoidance, even more attack surfaces will be opened. Additionally, as the software in vehicles becomes more complex, ever more frequent software updates must be issued, and each software update requires the opening of some port into the device. Reference [3] by Simon Burton *et al.* identifies four major points of attack on a car's lane departure warning system and reminds developers that security threats need to be included in a product's hazard and risk analysis and safety case.

Reference [4] by Harold Thimbleby captures the tension between usability, security, and safety very well:

> - *Usability is about making systems easy to use for everyone.*
> - *Security is about making systems easy to use for designated people and very hard (if not impossible) for everyone else.*
> - *Safety is about stopping the right people doing bad things. Good people make slips and errors; in a safe system those errors should not escalate to untoward incidents, harm or other safety issues.*
>
> *These simple and nicely contrasting definitions raise immediate tensions. A secure system may not be very usable. In turn this will encourage its users to find workarounds to make their daily use of the system more 'user friendly'. Thus a hospital system may end up with everyone sharing the same password, as this is much easier than remembering individual passwords and logging in and out repeatedly.*

It is easy to see how the lack of security can provide a safety hazard, but one problem with increasing security is that it almost always tends to reduce system performance. If a database is encrypted to maintain security, then it takes longer to access; if a strong password is enforced on the operator interface to a medical device, then it slows down access to that device.

Performance is not something that can always be given up lightly. The reverse-facing camera on a car is needed within a few hundred milliseconds of the car's systems being turned on. This may preclude the signature on the software image being checked at startup, thereby reducing the system security.

One other source of insecurity that must not be overlooked is the malicious developer with direct and legitimate access to the system's source code. This possibility needs to be recognized in the hazard and risk analysis as a potential risk and suitable mitigation put in place to detect it.

Performance/Reliability Balance

Increasing the reliability of a system typically means increased cost and reduced performance because some form of replication or diversification will have to be used — see Chapter 10.

Time replication (repeating the same computation on the same hardware, possibly using recovery blocks or coded processors) particularly reduces system performance, but other techniques to increase the probability of getting a correct system response, such as disabling cacheing on the processor, also directly hit performance.

Implementation Balance

There is one further balance that the architect or designer needs to consider. As the programmers implement the system they will need guidance on how to program. It is possible to program for high performance (fast, tightly-knit code), for testability, for ease of maintenance, for efficiency of the static analysis tools, for efficient runtime error detection, or for code readability.

These demands work against each other: High-performance code is likely to be less readable, less easily maintained, less likely to detect runtime errors, and less amenable to deep static analysis than is code specifically written with those characteristics in mind. Striving for high performance may particularly reduce the possibilities of runtime error detection. To facilitate this, data may be replicated, stored in diverse forms, or covered by a check-sum. Checking these at runtime will always reduce the performance of the system.

Programmers need to be aware of the priorities for each module they produce. The design of module A might specify that it is not time critical, and so the priorities when creating the code are for runtime error coverage, ease of testability, efficiency of the static analysis tools, and ease of maintenance, in that order. Module B might need to be programmed for speed, even if this means that more work will have to be spent testing and maintaining it, and even though fewer runtime errors will be detected.

Once these priorities have been set for each module, the resulting characteristics can be built into the system's failure model.

In practice, one problem may be getting the programmers to take notice of the allocated priorities — some programmers insist on coding for high performance, irrespective of the prioritization of other characteristics.

Summary

The design and implementation of any system, but particularly an embedded system designed for a safety-critical application, are subject to balances and tradeoffs. There is a long list of elements that must be balanced, including safety, performance, security, usability, availability and reliability. The analysis of these should be carried out explicitly and the decisions made should be recorded. This allows those decisions to be reviewed in the light of experience.

References

1. C. Hobbs and A. Azim, "Balancing Safety, Security, Functionality and Performance," in *2013 Safety Critical Systems Symposium*, SSS '13, (Bristol, UK), Safety-Critical Systems Club, 2013.
2. K. Koscher, A. Czeskis, F. Roesner, S. Patel, T. Kohno, S. Checkoway, D. McCoy, B. Kantor, D. Anderson, H. Shacham, and S. Savage, "Experimental Security Analysis of a Modern Automobile," in *Proceedings of the 2010 IEEE Symposium on Security and Privacy*, SP '10, (Washington, DC, USA), pp. 447–462, IEEE Computer Society, 2010.
3. S. Burton, J. Likkei, P. Vembar, and M. Wolf, "Automotive Functional Safety = Safety + Security," in *Proceedings of the First International Conference on Security of Internet of Things*, SecurIT '12, (New York, NY, USA), pp. 150–159, ACM, 2012.
4. H. Thimbleby, "Safety versus Security in Healthcare IT," in *Addressing Systems Safety Challenges, Proceedings of the 22nd Safety-Critical Systems Symposium* (C. Dale and T. Anderson, eds.), pp. 133–146, 2014.

Chapter 8

Error Detection and Handling

Law Number XVIII: It is very expensive to achieve high unreliability. It is not uncommon to increase the cost of an item by a factor of ten for each factor of ten degradation accomplished.

Norman R. Augustine

Why Detect Errors?

Given the fault→error→failure chain described on page 16, it can be useful in a running system to detect the error caused by the (undetected) fault and take appropriate action before it becomes a failure. Once the system is running, it is too late to detect faults, but errors often leave observable traces. By detecting the error, it may be possible to avoid a failure completely, but even if it is not, it may be possible to log details of the problem and inform the larger system into which the failing component is embedded before the failure occurs.

Some techniques for detecting an error, as listed in Table 4 of ISO 26262-6, are straightforward and likely to be applied to any software development — for example, functions performing range checks on their input parameters — while others, such as "plausibility" checking, are more subtle and are considered in this chapter.

Error Detection and the Standards

Table A.2 of IEC 61508-3 recommends the incorporation of *fault detection* into a software design. The associated description given in C.3.1 of IEC 61508-7 indicates that, using the terminology of "fault," "error," and "failure" described on page 16, the standard is actually addressing errors rather than faults.

ISO 26262 dedicates Table 4 in part 6 to various error detection mechanisms, recommending range and plausibility checks on input data and various forms of internal and external monitoring.

Anomaly Detection

Anomaly detection is the application of an algorithm to a series of events to decide whether one or more of the events is unexpected in the light of the history of the events up to that point. Humans are very good at this: One glance at Figure 8.1 is enough for a human to know that, even with the noisy signal, something strange happened around time 150. This section describes two different types of algorithm for detecting such anomalies and deciding *how* anomalous they are.

Two types of anomalies can be identified: *point anomalies,* such as occur in Figure 8.2 at time 308, where a few unexpected points occur and *contextual anomalies,* such as that in Figure 8.1 around time 150, where the points by themselves are not anomalous, but the pattern is.

The remainder of this chapter deals with detecting anomalies at runtime in systems. Anomaly detection can also be useful offline, during integration testing — see page 291.

Anomaly Detection and the Standards

While not listed explicitly, anomaly detection is a form of plausibility check recommended in Table 4 of ISO 26262-6. A series of measurements is made and, for each measurement, the question is asked, "how plausible is this value, given the history of previous values and the knowledge of the physics of the system?"

Both algorithms for anomaly detection described in the remainder of this section involve a form of unsupervised learning to create a baseline of "normal behavior" from which anomalies can be detected. It can be harder to justify the use of a learning rather than a static algorithm, because the functionality of the system is not predefined, depending instead on what the system learns. Such dynamic algorithms are discouraged by the standards.

This can be mitigated somewhat, at the cost of decreased flexibility, by letting the system learn in a controlled environment, perhaps during verification testing or from event logs gathered from the field. The resulting static algorithm can then be deployed.

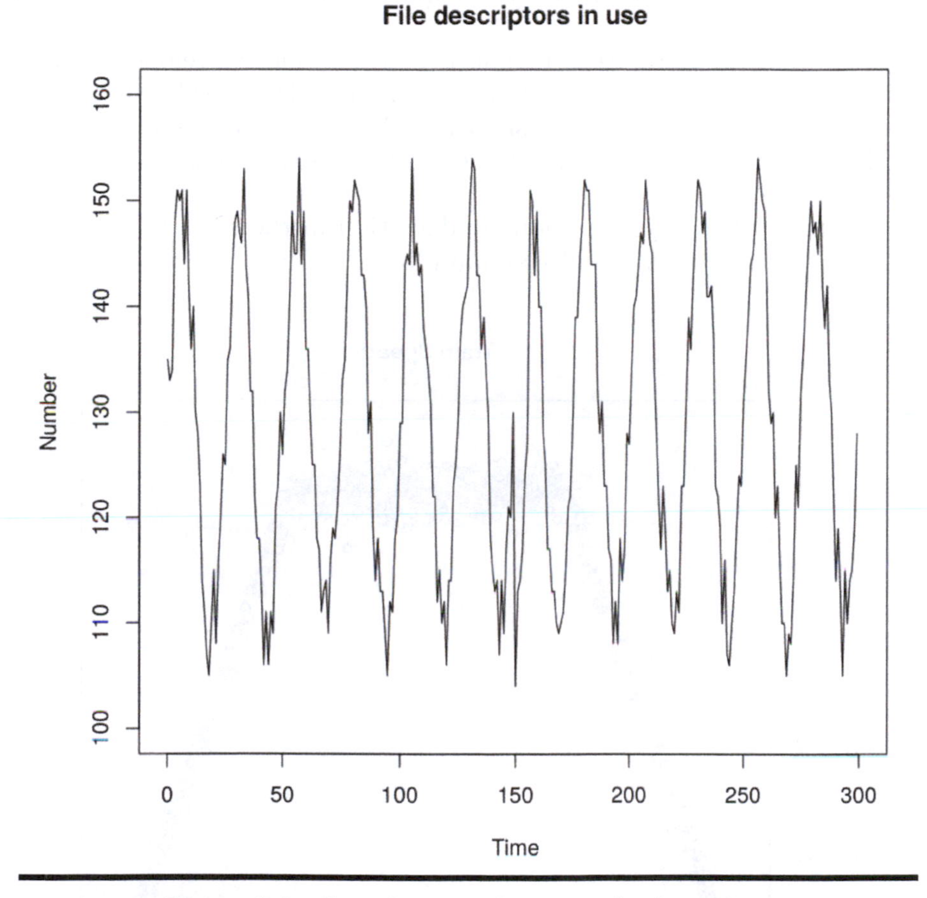

Figure 8.1 Sample trace for anomaly detection.

Algorithms for Anomaly Detection

Many techniques have been proposed for detecting anomalies, and the survey article, reference [1] by Varun Chandola *et al.*, lists several dozen.

Much of the research being carried out on anomaly detection is aimed toward the detection of malicious intrusions into networks by attackers and the examination of financial databases, looking for patterns of fraud. The results of this work have to be carefully considered to see whether those security-related techniques also apply to more general anomaly detection of the type required to detect faults becoming errors.

One promising area that I think is possibly underdeveloped is that of using wavelet transforms to detect anomalies. The wavelet transform of a time series, such as those shown in Figures 8.1 and 8.2, preserves both its spectral and, because the scale of the wavelet can be changed, its time aspects.

I consider two types of anomaly detection in the remainder of this chapter: Markov chains and Kalman filters.

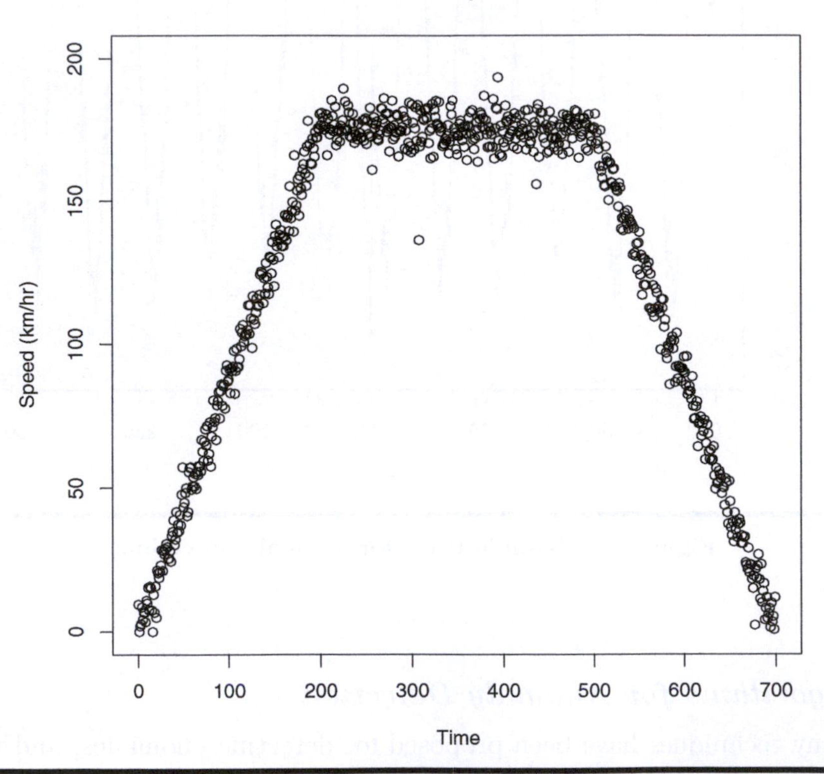

Figure 8.2 Sample data.

Examples

As examples of anomaly detection, we will use the fictitious samples illustrated in Figures 8.1 and 8.2.

The values in Figure 8.1 represent the exact number of resources in use over time. I have assumed that the resource is an open file descriptor, but it could be an amount of memory, a number of locked mutexes, or any other resource whose usage can be measured precisely. Clearly something odd happens around time $t = 150$ in Figure 8.1.

Figure 8.2 presents a different type of anomaly. The values are assumed to be the speed measurements from a forward-facing Doppler radar on a train. The train accelerates to around 170 km/hr, cruises at that speed for a while and then comes to a stop.

In contrast with Figure 8.1, where the values were exact counts, in Figure 8.2 there is some noise on the readings caused by the nature of the radar and there are some values that, to the eye, immediately appear to be anomalous: particularly those at $t = 308$ and $t = 437$. Figure 8.3 expands the section of Figure 8.2 around the $t = 308$ anomaly. Such a situation, where a sensor is providing inexact and possibly anomalous values, and where the main system needs to handle these, reflects the conditions aboard Qantas Flight 73, as described on page 280.

We can easily pick out the anomalies in Figures 8.1 and 8.2, because we can see not only what occurred before them, but also what happened afterwards. The questions we need to address for the running system are "while receiving the values from the radar in Figure 8.2 at time $t =$ 308, without the benefit of seeing into the future, is the reading of 136.5 km/hr that we have just received anomalous? If so, *how* anomalous?"

Anecdote 8 *When discussing anomalies, I often refer to the Swiss artist, Ursus Wehrli. Wehrli creates "TIDYING UP ART," such as parking lots with all the cars of the same color parked together, or a bowl of alphabet soup with the letters arranged alphabetically.*

The point is that we tend to think of an anomaly as something suddenly different; with Wehrli's examples, the anomaly is in the regularity: It is anomalous to have cars in a parking lot sorted by color or the pasta in a soup bowl sorted into alphabetical order.

The leakage of file descriptors described in Anecdote 9 is an example of a "regular" anomaly. The file descriptors were leaking extremely regularly, and this was anomalous.

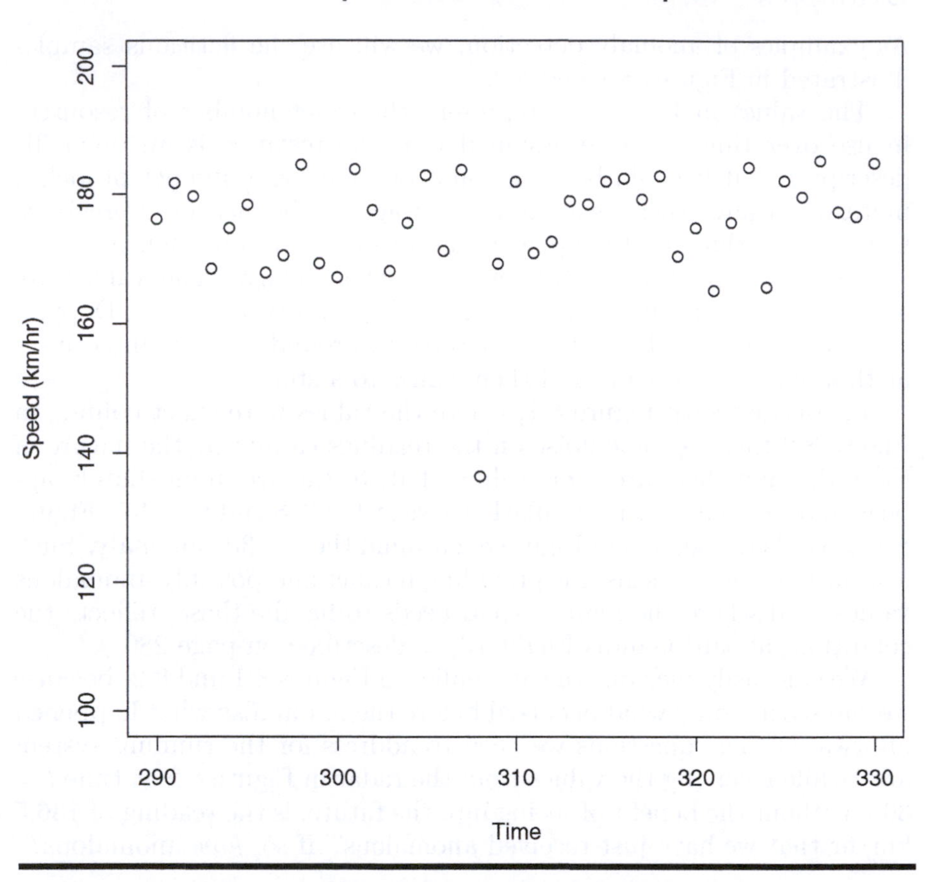

Figure 8.3 Expanded view of sample data.

Markov Chains

Andrey Markov

Markov chains and the Markov models described in Chapter 11 are named after Andrey Markov, a member of a Russian mathematical family, with his brother and son also making significant mathematical discoveries. He lived from 1856 to 1922 and made many important contributions to the theory of stochastic processes. His life was spent in academia although he was dismissed from his professorship at St. Petersburg University in 1908 for refusing to report to the authorities any revolutionary tendencies amongst his students.

System States

The first step in representing a system as a Markov chain is to itemise its possible states. For the system shown in Figure 8.1, an obvious but incorrect way of representing the state at any time is by the number of file descriptors in use. If 131 file descriptors are in use, then the system is in "state 131."

While this is simple, it unfortunately breaks the fundamental assumption of Markov chains — that the probability of being in state j at time $t + 1$ depends only on being in state i at time t. The history of how state i was reached is irrelevant. This is described more thoroughly in Chapter 11, but assume for the system in Figure 8.1 that the current number of file descriptors in use is 131 and that we want to predict how many will be in use at the next sampling time. To apply a Markov chain, we must assume that this will not depend on any earlier values. In Figure 8.1 this is not true, because the use of file descriptors is clearly cyclic, and the value of 131 can occur both on a downward and an upward part of the cycle. During a downward cycle, the probability of moving from 131 descriptors in use to 129 in use is much higher than it would be on an upward cycle.

To avoid this, our representation of the system state will have to include details of whether the system is on the downward or upward part of the cycle. The current system state could therefore be defined to consist of two parts: the number of file descriptors in use and whether, in the previous time interval, the number of descriptors in use went up or down. So the system state might be 131U or 131D. If the number of descriptors doesn't change, then we could apply a third suffix, but I have just assumed that the suffix is U.

In principle, it would then be possible to draw a diagram showing the possible transitions between states and their likelihood. From Figure 8.1, a transition between state 131U and state 133U seems likely, whereas a transition from state 131U to state 110D seems highly unlikely. I say "in principle" because the diagram would contain many states and have many arrows between states indicating potential state transitions. Using the learning data described in the next section, there would be about 100 states and 2014 arrows between them. Figure 11.2 on page 159 illustrates a much smaller Markov chain, and even that figure looks somewhat confusing.

In some systems, the system states themselves form a Markov chain; in other systems, it may occur that internally the system is operating in accordance with a Markov chain, but that these states are not directly visible. In the latter case, it may be possible to use the technique of a "hidden Markov model," whereby the characteristics that *are* visible can be used instead of the Markov states themselves.

Learning the Transition Probabilities

In order to decide when something is anomalous, we need to know what is normal. For a Markov chain, we build the probability of transitions between states by observing normal behavior, possibly during system testing, possibly during field trials. This allows us to build a matrix of transition probabilities.

Training consists of observing the system and determining what proportion of the transitions out of state i are moves to state j. For example, for the system in Figure 8.1, my program observed 10,000 readings and calculated the transition probabilities. It found that the transition probabilities out of state 131U were as shown in Table 8.1. Note that that table is arbitrarily divided into "upward" (more file descriptors used) and "downward" (fewer file descriptors used) for the sake of readability.

Table 8.1 Transitions from the state 131U.

Transitions Downwards		Transitions Upwards	
New State	**Probability**	**New State**	**Probability**
117D	0.0250	131U	0.0625
118D	0.0250	132U	0.0375
119D	0.0125	133U	0.0250
121D	0.0250	134U	0.0750
122D	0.0125	135U	0.0750
124D	0.0250	136U	0.0500
125D	0.0250	137U	0.1000
126D	0.0125	138U	0.0375
127D	0.0250	139U	0.0875
129D	0.0500	140U	0.0250
130D	0.0250	141U	0.0125
		142U	0.0250
		143U	0.0875
		144U	0.0375

If the system is in state 131U, the previous state change was "upwards": i.e., toward more file descriptor usage. From that state, the probability is greater (73.75% to 26.25%) that it will continue moving upward, perhaps to as many as 144 descriptors in use. At this stage in the analysis, it would perhaps be better to replace the U and D on each state with an indicator not just of the last state change, but of the average of the last few state changes.

More formally, during learning we build a matrix

$$P = \begin{pmatrix} p_{1,1} & p_{1,2} & \cdots & p_{1,N} \\ p_{2,1} & p_{2,2} & \cdots & p_{2,N} \\ \vdots & \vdots & \vdots & \vdots \\ p_{N,1} & p_{N,2} & \cdots & p_{N,N} \end{pmatrix} \tag{8.1}$$

where $p_{i,j}$ is the probability of moving to state j at time $t+1$ if it is in state i at time t. During learning,

$$p_{i,j} = \frac{N_{i,j}}{N_i} \tag{8.2}$$

where N_i is the number of times that the system left state i, and $N_{i,j}$ is the number of times that the system moved directly to state j from state i.

From a practical point of view, this matrix is likely to be very sparse (i.e., have many zero elements). It may therefore be better to store it as an associative array* rather than as a matrix.

Clearly, some of the elements of the state transition matrix will be zero. For example, during the learning phase on the file descriptors, the system never went directly from 131 to 148 file descriptors in use. Thus, the entry in P (Equation 8.1) corresponding to that transition will be zero. Having zeroes in the transition matrix will cause some problems during the analysis phase, and so it is useful to replace all the zeroes by very small numbers (e.g., 10^{-6}).

Looking for an Anomaly

Once we have determined what is "normal" during the learning phase, we can look for anomalies that occur during the operation of the system.

The objective is to look at a series of transitions in the system states and determine, given the transition probability matrix (Equation 8.1), how likely it is for that series to have occurred in normal operation.

---

* An array indexed by a value other than an integer. For example, `x['sValue']`, where the array x is indexed by a string. Some languages, such as Python, support associative arrays directly; in other languages, the construct may have to be emulated.

Given a "window" of P transitions, the likelihood is calculated as:

$$L = \prod_{t=1}^{P-1} P(X_{t-1}, X_t) \tag{8.3}$$

where $P(a, b)$ is the probability of transition from state a to state b. It can be seen from this equation why values of zero are replaced by small, but positive, values — any single zero would dominate the entire computation. Once this calculation has been performed, the window is moved along one time period and repeated.

Some experimentation is needed to determine the window size — the length of the sequence to be observed. Figure 8.4 displays the analysis on the number of file descriptors in use for a window size of 5; the vertical axis being a measure of how likely it is that the series of 5 states would have occurred during normal operation, without an anomaly being present.

It can be seen that the measure of likelihood of the sequence of transitions following time $t = 150$ is effectively zero: there is clearly an anomaly at that time. The computation also finds anomalies around time 200 and 260. These are not as immediately obvious when looking at Figure 8.1.

Kalman Filters

History of Kalman Filters

Kalman filters are named after Rudolf E. Kálmán, a Hungarian mathematician who emigrated to the USA in 1943 at the age of 13. He introduced the Kalman filter in reference [2], which was published in 1960. It is often said that Thorvald Nicolai Thiele in the 19th century and Peter Swerling in the 20th century devised the filter earlier than Kálmán, and some threads can even be traced back to Gauss.

Summary of the Discrete Kalman Algorithm

A Kalman filter is a predictor/corrector algorithm, as illustrated in Figure 8.5. At time T, the algorithm predicts the value that will be received from the sensors at time $T + 1$, and, when the actual reading is received, it adjusts the predictor to make it more accurate next time. It then makes a new prediction and this process continues indefinitely. Predicting what the next reading will be is useful for anomaly detection because, once the algorithm has settled down and is accurate in its predictions, a major difference between the predicted and actual values represents a potential anomaly.

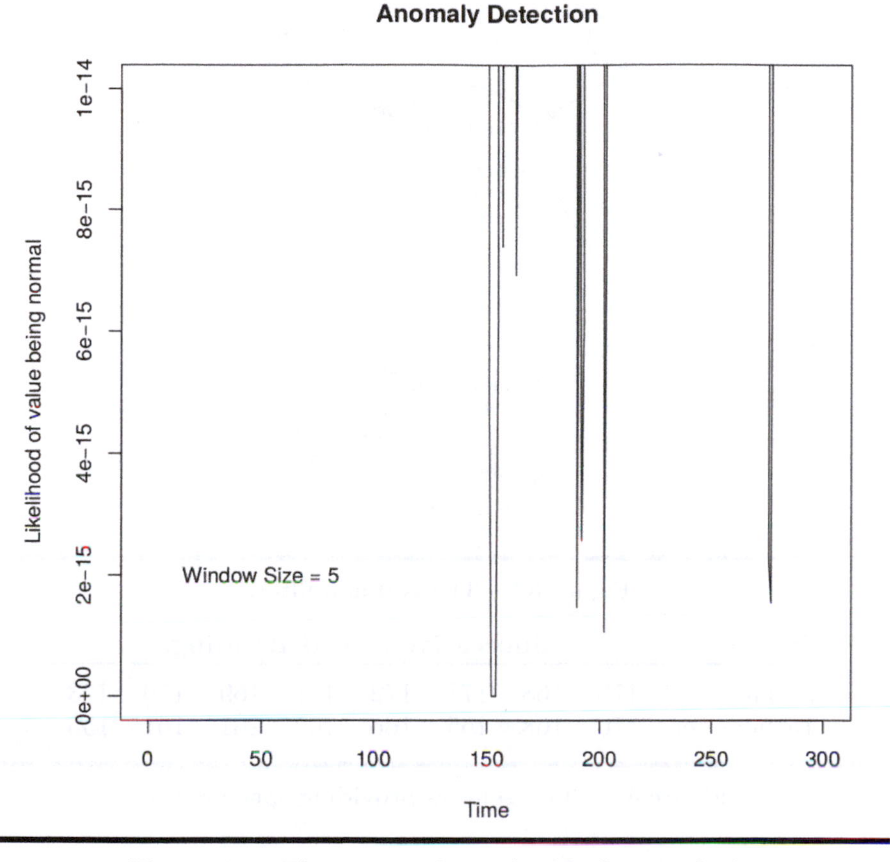

Figure 8.4 The output from the Markov analysis.

The algorithm can look mathematically intimidating,* but for most systems is very simple to program, particularly, as is often the case, when the values are one-dimensional. The train speeds given in Figures 8.2 and 8.3 are single numbers sampled at each timer tick. If there were two sources of speed input (e.g., Doppler radar and counting of wheel rotations), then a two-dimensional filter could be used to detect anomalies between the readings.

For example, in Figure 8.6, neither sensor by itself appears to be anomalous; what is anomalous is that one sensor is detecting a deceleration, while the other is not.

* Refer to appendix C for a list of the common symbols used.

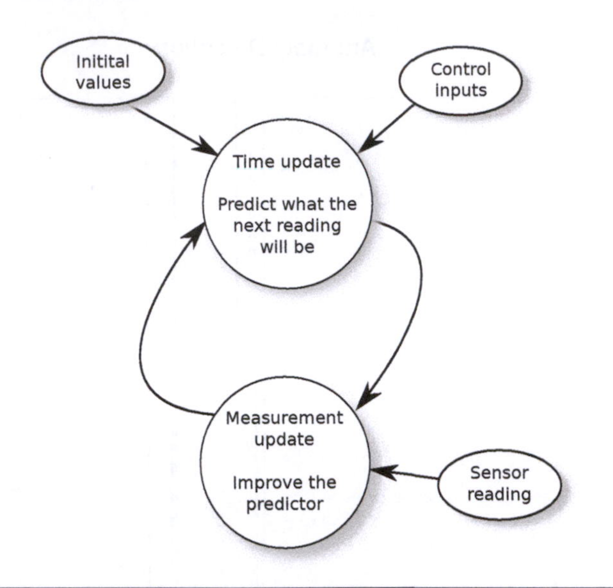

Figure 8.5 The Kalman filter.

Sensor	Successive Speed Readings							
Radar	170	168	171	172	171	169	170	173
Tachometer	170	168	165	166	162	161	161	155

Figure 8.6 Two sensors providing speed data.

Details of the Kalman Algorithm

Reference [3] by Greg Welch and Gary Bishop provides a gentle introduction to the algorithm. In the following, I use the notation from that paper, but simplify the explanation somewhat.

Assume that the state of the system can be described by a vector, $\vec{x}$ of n values (for the Doppler radar measurements in Figure 8.2, $n = 1$; for the radar and tachometer values in the table above, $n = 2$). A Kalman filter predicts the future value of $\vec{x}$ if the system can be described at the time k by an equation of the form

$$\vec{x}_k = A\vec{x}_{k-1} + B\vec{u}_{k-1} + \vec{w}_{k-1} \tag{8.4}$$

and measurements (sensor values) it receives measurements (sensor values)

$$\vec{z}_k = H\vec{x}_k + \vec{v}_k \tag{8.5}$$

The precise meaning of the terms are given below, but, in summary,

Equation 8.4 says that the new value of $\vec{x}$ at time k is some multiple of what $\vec{x}$ was at the previous point in time, plus some control input (e.g., the driver applying the brakes), plus some noise that is built into our system.

Equation 8.5 says that the new measurement ($\vec{z}$) is a multiple of what we predict $\vec{x}$ to be, plus some noise (error) in the measurement.

More precisely:

- $\vec{x}_k$ is the state of the system at time k. This is unknown to the software, which knows only the noisy measurements of $\vec{x}$ that it is receiving from the sensors. Generally, $\vec{x}_k$ is a vector incorporating all of the values of interest (e.g., distance, speed, acceleration). In the simplest case, as with the train speed in Figure 8.2, $\vec{x}_k$ could be a single number.
- A is an $n \times n$ matrix that relates $\vec{x}_{k+1}$ to $\vec{x}_k$ according to the physics of the problem, assuming that there is no noise on the sensor readings.

  An example for a train moving along a track would be that the new distance travelled is the previous distance travelled, plus the average speed multiplied by the length of the time interval. In most applications, A does not change in time and, if $\vec{x}$ is a single number, then A becomes a single number (or, more correctly, a 1×1 matrix), normally equal to 1.
- Generally the condition of the system will vary according to changes in its control inputs, and we know from the system design what these changes should be. We can tell the Kalman filter about these so that it doesn't see the changes as anomalous.

  For example, if the system has information about the position of the driver's control lever, then this could be incorporated into the equation — if the driver reduces power, then we would expect the train to slow down. This is helpful information for the algorithm when it is predicting the next value of $\vec{x}$. This value could also incorporate environmental factors; for example, the train starting to move uphill.

  B is an $n \times l$ matrix that relates one or more control inputs to the new values of $\vec{x}$.
- $\vec{u}_k$ is a vector of length l that relates the control input specified by B to the expected change in $\vec{x}_{k-1}$.
- $\vec{w}_k$ is the noise in the process itself (not the noise in the measurements). For many systems, the process itself is not noisy — only our measurements of it are noisy. In those cases $\vec{w}_k$ can be set to $(0, 0, 0, \ldots, 0)$.
- $\vec{z}_k \in \mathbb{R}^m$ is the measurement taken from the m sensors at (dis-

crete) time k. Again, if there is only a single sensor, as in the case of the train speed in Figure 8.2, then $\vec{z}_k$ becomes just one number.

- H is an $m \times n$ matrix that relates the actual system state to a measurement. If there is a single sensor, then H would be a single number, normally with value 1.
- $\vec{v}_k$ is the noise in the measurements. This is assumed to be independent of $\vec{w}_k$.

Limitations

As illustrated in Equation 8.4, it must be possible to express the system values at time k to the system values at time $k-1$ in a linear manner. This may be inappropriate for some internal anomalies, such as number of file descriptors in use, where there is no physical law governing the system.

The Kalman filter only works perfectly if the noise values ($\vec{w}_k$ and $\vec{v}_k$) are independent of each other and are Gaussian and have mean values of 0. In what follows, I assume that the covariance matrices are Q and R respectively. For the detection of internal, rather than sensor-based, anomalies, both $\vec{w}_k$ and $\vec{v}_k$ are typically 0, and so this limitation does not arise.

Prediction and Correction

Given the symbols listed above, the prediction and correction equations are as follows:

Prediction

$$\vec{x}_k = A\vec{x}_{k-1} + Bu_{k-1} \tag{8.6}$$

$$P_k = AP_{k-1}A^T + Q \tag{8.7}$$

Equation 8.6 simply says that the best prediction we can make about the next value of $\vec{x}$ is to apply the A matrix to the previous value of $\vec{x}$ and then make an adjustment for any control input.

In Equation 8.7, P is the covariance matrix of the errors and can be initialized to any nonzero value as it will iterate to a correct value over time. Q is the process noise covariance and can normally be set to something very small (a value of 0 should be avoided, but for a one-dimensional model, a value such as $Q = 10^{-5}$ would not be unreasonable).

Correction

$$K_k = P_k^* H^T (H P_k^* H^T + R)^{-1} \tag{8.8}$$

$$\vec{x}_k = \vec{x}_k^* + K_k (z_k - H \vec{x}_k^*) \tag{8.9}$$

$$P_k = (I - K_k H) P_k^* \tag{8.10}$$

where the stars indicate the predicted values for time k.

Anomaly Detection with Kalman Filters

A Kalman filter can be used to detect anomalous values read from the sensors. A predicted value ($\vec{x}_k^*$) for the next reading can be compared with the actual reading when it arrives and the difference

$$\vec{v}_k = \vec{z}_k - H \vec{x}_k^* \tag{8.11}$$

can be calculated (see Equation 8.9).

This represents in some way the "unexpectedness" of the sensor value, and

$$e^2 = v_k P_k v_k^T \tag{8.12}$$

is distributed as χ^2 with the number of degrees of freedom being the number of measurements. Any desired level of confidence can therefore be placed in the degree by which a particular reading is anomalous.

Anomaly Probabilities for the Train Example

Figure 8.8 illustrates the probability of anomaly of the values given in Figure 8.2. As has been stated above, this is a particularly simple application of the Kalman filter because it is one-dimensional because the only value of interest is the train's speed. This means that all the vectors and matrices deflate into simple numbers, and the code to perform the algorithm is extremely simple. The predictor, for example, reduces effectively to two lines of Python code calculating Equations 8.6 and 8.7 as shown in Figure 8.7.

I have arbitrarily set the action threshold to 0.5 in Figure 8.8, but that value would be application-specific. Note that, once the system has settled down and the filter is able to make reasonable predictions about the next value, the probability of anomaly remains generally low, apart from the value given at time $t = 308$, which we know from Figure

```
def kf_predict(x, p, a, q, b, u):
    '''
    Perform the prediction step
    Input x : mean state estimate of the previous step
          p : state covariance of previous step
          a : transition "matrix"
          q : process noise covariance "matrix"
          b : input effect "matrix"
          u : control input
    Output x : new mean state estimate
           p : new state covariance
    '''

    newX = (a * x) + (b * u)
    newP = (a * p * a) + q

    return(newX, newP)
```

Figure 8.7 Kalman predictor code.

8.2 to be anomalous. It is interesting that the other value that appears to be anomalous in Figure 8.2, when $t = 437$, appears as a spike in Figure 8.8, but at a much lower level than that of $t = 308$.

Both the threshold at which action needs to be taken and the choice of action are system-dependent. Perhaps in this case, the sensor value received at $t = 308$ will be discarded and replaced by a replica of the value at $t = 307$.

Rejuvenation

We now turn from detecting errors to one method of handling them. As stated above, our purpose is to prevent an error from becoming a failure or, if that is not possible, to prepare for the failure in as controlled a manner as possible.

Rejuvenation is one means of breaking the link between error and failure in a controlled manner. The term "rejuvenation" is just the formal way of saying, "press the RESET button from time to time". Most people who work with computers know that turning them off and on again is a cure for many software maladies, and it seems reasonable to apply this principle, where possible, to embedded systems too. Obviously, there are some devices, for example an embedded heart

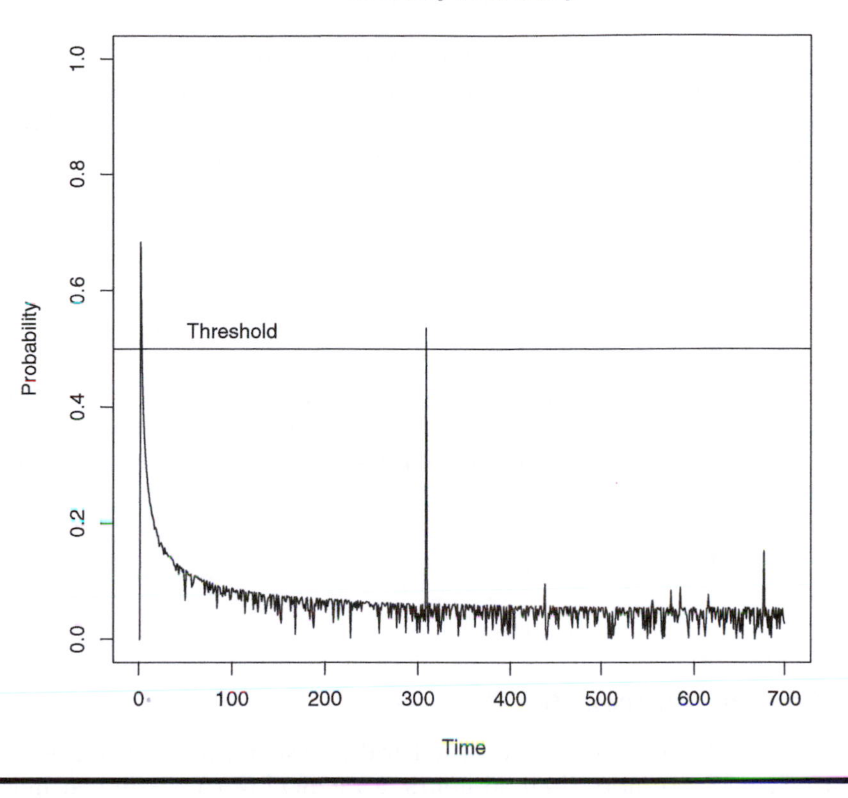

Figure 8.8 Anomaly probability.

pacemaker, where rejuvenation may not be practical.

The purpose behind rejuvenation is to intercept and remove errors before they become failures. Faults in the code may have caused a number of errors — zombie threads, data fragmentation and Heisenbug corruptions of memory, resource leakage, etc. — that have occurred but have not yet caused failures.

It is interesting to note that the problem aboard Qantas Flight 72 described on page 280 was solved by rejuvenation. According to the official report by the Australian Transport Safety Bureau:

> *Once the failure mode started, the ADIRU's abnormal behaviour continued until the unit was shut down. After its power was cycled (turned OFF and ON), the unit performed normally.*

Anecdote 9 *I was once working on a mission-critical system (luckily, not safety-critical) in the City of London's financial district. We had installed a highly dependable system based on virtual synchrony; see page 141. Once some early teething troubles had been overcome, we were proud of the 24 hours per day, 365 days per year operation of the system.*

After three years, it failed. Investigation showed that there was a fault in the code which was opening a file regularly but not closing it. The error was that file descriptors were being used up. The failure occurred when they were exhausted.

Had we deployed some form of rejuvenation (say, a reboot of the system once per year), the fault would still have been present and the errors would still be occurring, but the failure would not have happened.

When to Rejuvenate

Most embedded systems have a natural operating periodicity. An aircraft flies for no more than 36 hours, a medical device is in continuous use for no more than a month, a car is driven for no more than a few hours, etc. Given this periodicity, it is unclear why a reboot should not be built into the operating requirements at these convenient times. I think that in part this is due to a designers' natural tendencies to be heroic, demanding a periodic rebooting of the system is considered by many to be failure in some way.

What Does Rejuvenation Do?

Viewed from an availability analyst's point of view, rejuvenation is a form of periodic replacement, such as occurs in hardware maintenance — the piston engine on a light aircraft, for example, must be replaced every 1800 hours rather than waiting for it to fail before repair.

Figure 8.9 illustrates the effect of rejuvenation by comparing two Markov models (see Chapter 11). In the left-most diagram, where there is no rejuvenation, the system moves from its fresh state to a stale state (that is, a state where errors are emerging, but no failure has occurred) at a rate of r_1 per hour. Once in the stale state, there is a rate, r_2, at which the system moves into its failure state. The rate r_3

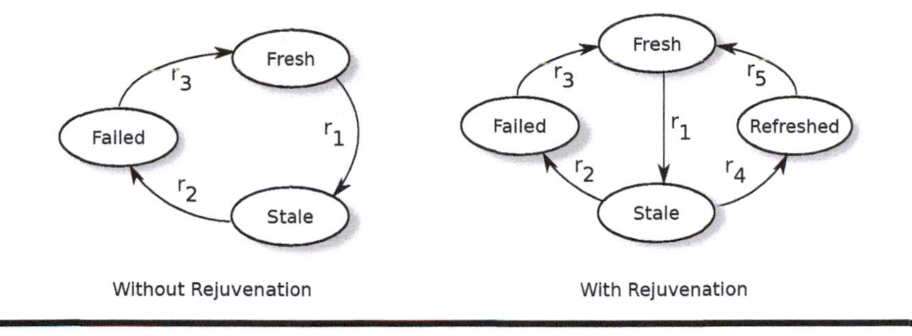

Figure 8.9 The effect of rejuvenation.

then represents the repair rate.

The right-most diagram in Figure 8.9 shows how rejuvenation affects the system. When the system is stale, rejuvenation occurs at a rate r_4, and this can avoid the move to the failed state.

Choosing r_4

The main problems associated with rejuvenation are determining the correct rejuvenation rate (r_4) and reducing the time required to complete the rejuvenation (by maximising r_5). If r_4 is too large, representing very frequent pressing of the reset button, then system availability drops. If r_4 is too small, the technique has limited use, because the r_2 transition can dominate when the system is stale, causing a failure and again impacting the availability of the system.

There are various ways of selecting r_4. If the system has an operating time, $\frac{1}{r}$ that is such that $\frac{1}{r} \gg \frac{1}{r_2}$, then rejuvenation can take place during its maintenance period. As described above, this may be valid for systems in aircraft, cars, railway trains, medical devices, and other systems regularly taken out of service. In this case, it is essential that the maximum rejuvenation interval be explicit in the safety manual and be included as part of the claim in the safety case:

> *We claim that the system does X as long as it is operated for no more than Y hours between resets.*

Of course, there are systems that are not taken out of service sufficiently often for rejuvenation to occur during routine maintenance, for example, a drill head pressure monitor on the sea bed, or a device monitoring the vibrations from a continuously operating electrical generator, and these need to be treated differently.

In these cases, the rejuvenation may need to be triggered by an anomaly detection system; see page 98. For example, it would be possible to trigger rejuvenation if the number of open file descriptors passed a predefined threshold or the amount of free memory dropped below a threshold.

Rejuvenation in Replicated Systems

In cases where a replicated or diverse design has been employed, particularly a hot/cold standby model of the type shown in Figure 10.2 on page 134, rejuvenation may be useful not only to cleanse the operating (hot) subsystem, but also to exercise the changeover mechanism and ensure that the cold (non operating) subsystem has not failed silently.* If this is done before the hot subsystem has failed, a rapid switch may still be possible back to the original subsystem if it is found that the standby has failed silently.

In this model, rejuvenation may be triggered by time (invoking the switch-over every X hours) or whenever an anomaly is detected in the hot (operating) subsystem. Kalyanaraman Vaidyanathan *et al.* describe a way of designing such a system in reference [4], using a special type of stochastic Petri net (see page 199). Although the analysis in reference [4] was carried out on a network of computers, the results are also valid for a replicated or diverse embedded system. Note that the reference uses the term "prediction-based rejuvenation" for what this book terms "anomaly-triggered rejuvenation."

Rejuvenation and the Standards

IEC 61508-2, the hardware section of IEC 61508, refers to the concept of "proof testing" for hardware. This is defined in IEC 61508-4 as follows:

> *proof test: periodic test performed to detect dangerous hidden failures in a safety-related system so that, if necessary, a repair can restore the system to an "as new" condition or as close as practical to this condition.*

This is the hardware equivalent of the rejuvenation described above, and reference [5] by Stuart Main and Ron Bell provides an insight into the complexities of selecting, in the terms of Figure 8.9, a suitable value for r_4. The reference also covers the case where the proof test

* A subsystem is said to have failed silently if its failure has not been noticed by the system.

is imperfect — in our case when, because of the system design, the rejuvenation cannot be total.

Recovery Blocks

Another particularly simple (and effective) technique for detecting an error and recovering is that of "recovery blocks." The term "block" reflects the origin of the method back in the ALGOL language that was in use in the 1970s. Reference [6] by Jim Horning *et al.* provides one of the earliest references to the technique (1974), and Brian Randell and Jie Xu's 1994 paper, reference [7], describes how the concept evolved until that time. The work is generally believed to have arisen initially as a response to the "software crisis" identified at the 1968 and 1969 NATO Software Engineering Conferences.

Fallacy 5 *I have referred early to Laurent Bossavit's book on the leprechauns of software engineering (reference [8]). Chapter 8 of that reference contains a very strong argument that the well-known "fact" that the NATO conferences were convened in response to rising recognition of the software crisis is, in fact, completely untrue. Apparently, none of the illustrious group of people at the conferences which included a who's-who of software names (Bauer, Dijkstra, Hoare, Naur, Randell, Wirth, and many, many others), ever referred to a "crisis," and the term did not start to be used until five years after the conferences.*

Recovery blocks are explicitly mentioned in Table 5 of ISO 26262-6 as a "Mechanism for error handling at the software architectural level."

The basic pattern of a recovery block is that, instead of calling a particular function to carry out a computation, we instead structure the code as follows:

```
ensure <acceptance test> by <function1>
else by <function2>
...
else by <functionN>
else error
```

The acceptance test is a routine that evaluates the result returned by the function and indicates whether it is acceptable or not in the system context.

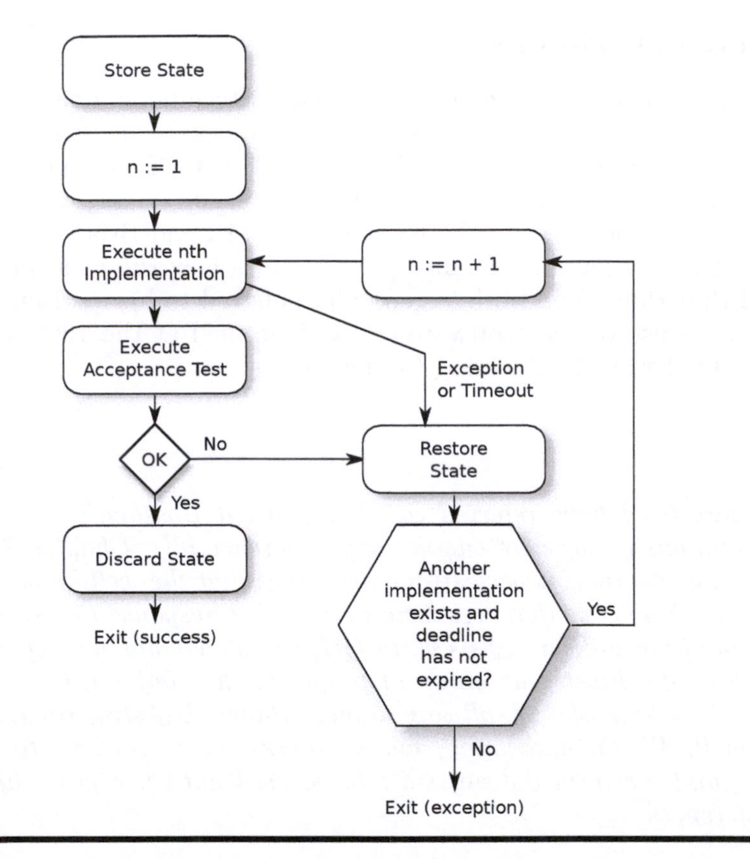

Figure 8.10 Recovery block with backward error recovery.

The flow of execution is illustrated in Figure 8.10, where recovery blocks are combined with backward error recovery; see page 19 for a description of the difference between backward and forward error recovery. The process is that function1 is executed, and the acceptance test is invoked on the result. If the result is found acceptable, then progress continues with the next program block. If the result is not acceptable, then function2 is invoked and its returned value is checked by the acceptance test. Eventually, either a result is obtained that is acceptable, or all the functions have been tried and an error exit is taken — perhaps moving the system to its design safe state; see page 123.

There are many variants of this basic technique, which can be thought of as a version of N-version programming, as described on page 139. One variant stems from the observation that the different implementations of the function do not need to calculate identical results, they only need to return *acceptable* results. So `function1` might try to handle the input parameters fully, whereas `function2` might perform a simpler computation giving a more approximate answer.

A second observation arises from the fact that most faults remaining in shipped code are likely to be Heisenbugs. If this is the case then the pattern can be simplified:

```
ensure <acceptance test>
by <function1>
else by <function1>
else error
```

Invoking the same function (here, `function1`) again if the first result is unacceptable is likely to lead to a acceptable result. By definition, the precise conditions needed to provoke the Heisenbug are unlikely to recur.

As an example of using recovery blocks with the same routine invoked twice, consider the 2-threaded program in Figure 5.4 on page 70. On my computer, this program fails every few hundred million times it is executed. Executing the program continuously 1000 times per second means that it will crash approximately every week. Catching that crash and simply rerunning the same program once will increase the time between crashes to something like 8 million years.

One disadvantage of recovery blocks is that any side effects (system changes) made by a routine whose output is eventually rejected by the acceptance routine must be undone before the computation is repeated. These side effects include what reference [7] terms "information smuggling," when the routine changes data structures outside the group of programs engaged in the processing or even dynamically creates new processes.

Recovery blocks can also provide some level of protection against security attacks, because the attacker has to corrupt both the implementation *and* the acceptance test before a bad value can be silently generated. In the second (2000) Turing Lecture, Brian Randell reported on an experiment his group performed with a system implemented with recovery blocks. Users were encouraged to make arbitrary changes to the code of various implementations. After a short period, the system had been honed to such a point that no amount of modification caused errors in its operation.

Unfortunately, no modern programming language supports recovery blocks directly (although D comes close), but the necessary constructions can be provided in almost any language through macros or meta-programming.

A Note on the Diverse Monitor

The diverse monitor pattern could reasonably have been included in this chapter as a means of error detection (that is how it is listed in Table 4 of ISO 26262-6) or in the chapter on diversification because it is a form of diverse design. In fact I have included it in Chapter 10 on page 147 as a form of replication.

Summary

Once the system is deployed and is running in the field, it is too late to detect faults. However, we can break the chain from fault to error to failure by detecting the errors as they arise and taking "appropriate" action to avoid the failure.

Anomaly detection is one mechanism for detecting the presence of errors. Rejuvenation is one means of avoiding failure (in the case of a replicated system), or at least of causing the failure to occur in a controlled manner, possibly at a time when it is convenient for the system to handle it. Recovery blocks allow an error to be detected and possibly corrected before it causes a failure.

References

1. V. Chandola, A. Banerjee, and V. Kumar, "Anomaly detection: A survey," *ACM Comput. Surv.*, vol. 41, pp. 15:1–15:58, July 2009.
2. R. E. Kalman, "A New Approach to Linear Filtering and Prediction Problems," *Journal of Fluids Engineering*, vol. 82, 1960.
3. G. Welch and G. Bishop, "An Introduction to the Kalman Filter," tech. rep., University of North Carolina at Chapel Hill, Chapel Hill, NC, USA, 1995.
4. K. Vaidyanathan, R. E. Harper, S. W. Hunter, and K. S. Trivedi, "Analysis and implementation of software rejuvenation in cluster systems," in *Proceedings of the ACM SIGMETRICS 2001 Conference on Measurement and Modeling of Computer Systems*, pp. 62–71, ACM Press, 2001.
5. S. Main and R. Bell, "Proof Testing ... the Challenges," in *Proceedings of the 20th Safety-Critical Systems Symposium*, SSS'12, Springer-Verlag, 2012.

6. J. J. Horning, H. C. Lauer, P. M. Melliar, and B. Randell, "A Program Structure For Error Detection And Recovery," in *Lecture Notes in Computer Science*, pp. 171–187, Springer-Verlag, 1974.
7. B. Randell and J. Xu, "The evolution of the recovery block concept," in *In Software Fault Tolerance*, pp. 1–22, John Wiley & Sons Ltd, 1994.
8. L. Bossavit, *The Leprechauns of Software Engineering: How Folklaw Turns into Fact and What to Do about It.* Leanpub, 2013.

Chapter 9

Expecting the Unexpected

To expect the unexpected shows a thoroughly modern intellect.

Oscar Wilde

This chapter, which discusses the system's design safe state, is really a continuation of the previous chapter on error detection. In the sense used in this chapter, the "error" is in the design when the system meets a condition that was not anticipated during the design phase.

Design Safe State

Defining the Design Safe State

For a system with safety implications, it is essential to acknowledge that it will sometimes encounter conditions that it was not designed to handle. These conditions may occur because of random corruption caused by hardware or software (Heisenbugs) or may simply be a circumstance unforeseen in the design that leads to a deadlock or livelock. The behavior of the system under these conditions must be defined and must be included in the accompanying safety manual so that the designer of any larger system into which this one is incorporated can detect the condition and take the appropriate action.

The action to be taken is to move, if possible, to the system's currently applicable "design safe state" defined during the system design.

For some systems, the design safe state may be obvious and unchanging over time. For example, traffic lights at a busy road intersection might all be set to red, and a system that controls the brakes on a train might always apply the brakes.

For other systems, the design safe state may not be so obvious and may vary from situation to situation. For a medical device that dispenses drugs, is it safer to continue drug adminstration when the monitoring system meets an unexpected condition or to stop the drug flow? In the first case the wrong dosage might be administered; in the second, the lack of drug may harm the patient.

For a car's instrument cluster, if the software in the graphics system meets an unanticipated situation, is it best to blank the screen, to freeze the information that is on the screen, or to display sane but incorrect information?

Anecdote 10 *I was once driving a car at speed on a very lonely road on a dark night when the car's battery exploded, disabling the power steering and braking and extinguishing the lighting of the analogue dashboard. It was very disconcerting suddenly to have no indication of my speed. Experiments in a simulator would presumably have to be carried out to determine the safest design safe state for an instrument cluster under different driving conditions.*

As another example, take a system that controls the braking of a car, perhaps part of a collision detection system. When the car is travelling at 10 km/hr on a deserted side road, a suitable design safe state might be for the failing system to apply the brakes. At 120 km/hr on a highway, suddenly applying the brakes might be exactly the wrong thing to do.

Whatever the design safe state, it must have three characteristics:

1. It must be externally visible in an unambiguous manner.
2. It must be entered within a predefined, and published, time of the condition being detected.
3. It must be documented in the safety manual so that a designer using the component can incorporate it into a larger design.

It is also useful if entering the design safe state causes some form of diagnostics to be stored so that the condition can be investigated. In some systems, particularly medical devices, such diagnostic collection may be mandatory.

However well the design safe state is defined, it must also be accepted that it will sometimes be impossible for the system to reach it: The internal corruption that has caused the unexpected condition may be so severe as to prevent the system from taking any form of coherent action. There is always some level of hardware that must behave properly to allow even a simple software process to execute.

Knowing the design safe states of components allows the designer who incorporates those components into a larger system to plan accordingly. If moving into the design safe state causes the system to discard input already received and assumed processed, then this must also be defined in the safety manual so that the designer of the larger system can handle the condition.

Design Safe State and the Standards

Section 8 of ISO 26262-3 addresses the system's safe state including the statement that, "If a safe state cannot be reached by a transition within an acceptable time interval, an emergency operation shall be specified." This means that the time taken to reach the safe state needs to be estimated or measured and made available in the safety manual. Many other parts of ISO 26262 also specify requirements related to the design safe state. In particular section 6 of part 4 reminds designers that the system must not only be able to move to its safe state, but mechanisms must also be in place to hold it there until external action is taken.

IEC 61508 also has many references to the design safe state and Annex D of part 3, which defines the contents of the safety manual, explicitly states that the safety manual must include the:

> *Design safe state: In certain circumstances, upon controlled failure of the system application, the element may revert to a design safe state. In such circumstances, the precise definition of design safe state should be specified for consideration by the integrator.*

Design Safe State and Dangerous Failure

The purpose of the design safe state is to provide the designer with a way of handling unexpected conditions. The question is often asked whether moving to the design safe state constitutes a dangerous failure and should therefore be considered in the calculation of the failure rate for IEC 61508's safety integrity level (SIL).

To meet the requirements of SIL 3, for example, a dangerous failure rate of $< 10^{-7}$ failures per hour of operation has to be demonstrated. If a system is so poorly designed that, on average, it meets an unexpected condition every 1000 hours, but always detects that condition and moves to its design safe state, then can that system be considered to meet the requirements of SIL 3?

This is a question of the usefulness/safety balance discussed on page 91, and certainly it can be argued that, almost by definition, a move to the design safe state is not a dangerous failure. However, I have tried to include it where possible in the IEC 61508 calculation because, even though the larger system should be designed to handle a component moving to its design safe state, such moves will always place stress on the larger system and move it closer to a dangerous condition.

Recovery

Recovery Techniques

When an unanticipated condition occurs, there are two possible actions: to attempt to recover, for example, by using recovery blocks as described on page 117, or to move directly to the design safe state.

Software-based recovery often takes the shape of executing scripts from a "recovery tree." These have the form:

> If process X reports that it has met an unexpected situation, then kill processes X, Y, and Z. Then restart Z, wait 200ms, restart Y, wait until flag A has been set, and then restart X.

The intent is to try to preserve as much of the state of the system as possible, to restrict the corruption that could be caused by invalid information being transferred to another part of the system, and then to restart and resynchronize the affected subsystems with as little system-level disruption as possible.

Do You Want to Recover?

By definition, any attempt to recover from an unanticipated situation means executing an algorithm that has not been tested, and this execution will happen at a time when the system state is ill-defined. For many systems it can be argued that executing untested code that implements an untested algorithm on a system whose state is ill-defined and certainly unexpected, is not suitable for a safety-critical device.

It also requires some level of co-operation from the process that has detected the inconsistency. The level of co-operation varies from very small (perhaps just avoiding interfering with the recovery software) to significant (reporting details of the problem to the recovery software). At the time that this co-operation is required, the failing process is in an ill-defined state, a state that the designers did not foresee.

Quite often such heroic recovery attempts to avoid moving to the design safe state are only *partially* successful, leaving the system in a worse condition than it might have been in before.

It is accepted that there are systems where recovery must be attempted, but in many cases, it may be better to move immediately into the design safe state when an inconsistency is detected.

Crash-Only Model

One model of moving to the design safe state is the "crash-only" model as described in reference [1] by George Candea and Armando Fox.

The crash-only model is self-explanatory — on detecting an unanticipated condition, the system crashes and stops any form of further processing. Crash-only has several advantages:

- It relies on only a small part of the system continuing to operate; recovery often depends on most of the system continuing to operate.
- It defines simple macroscopic behavior with few externally-visible states.
- It reduces outage time by eliminating almost all of the time associated with an orderly shutdown — although it may still be advantageous to store some information about the event.
- It simplifies the failure model by reducing the size of the recovery state table. The crash-only paradigm coerces the system into a known state without attempting to shut down cleanly. This substantially reduces the complexity of the code.
- It simplifies testing by reducing the failure combinations that need verification.

It may be possible to combine the crash-only model with fast reboot (see, for example, reference [2]) to minimize the time that the system is unavailable. However, this has to be closely controlled to prevent continuous reboot in the face of a solid hardware error.

Anticipation of the Unexpected by the Example Companies

Chapter 4 introduced our two fictitious companies. Both of these need to define a design safe state for their system.

Beta Component Incorporated (BCI) is supplying Alpha Device Corporation (ADC) with an operating system for ADC's device, and ADC has specified in the contract that the operating system be certified as being suitable for use in ASIL C devices in accordance with ISO 26262.

This means that BCI will have to supply ADC with a safety manual for the operating system, describing how it can be used safely and defining the design safe state. As the operating system is simply a component of ADC's device,* it can have no knowledge of how the full device is operating in the car, and therefore cannot adapt its design safe state to the external situation. The crash-only model is therefore likely to be a suitable design safe state — when encountering a situation that it was not designed to handle, it will crash in an externally visible manner, possibly after storing diagnostic information.

If BCI's operating system is a monolithic** one, then the safety manual will probably also place limitations on how the included subsystems may be used. A subsystem, such as a TCP/IP communications stack, is an extremely complex component, impossible to test thoroughly. If the failure of such a component could affect the behavior of the operating system kernel, then limitations would probably be placed on its use.

If BCI's operating system incorporates a micro-kernel, where these complex subsystems cannot affect the kernel, then the restrictions in the safety manual are likely to be fewer.

ADC has a more complex analysis to perform to determine the design safe state for its device. We have not defined what ADC's device actually does, but even the simplest device can raise questions about what state can be considered safe. Consider the controller for a car's electric windows. When a user presses a button, the window should close, detecting the presence of a child's hand (or neck), and refusing to continue closing should that sort of resistance be met. The first thought is that the design safe state should be to open the window, but that may be dangerous in some situations: e.g., if the car has been driven

* Actually, as described on page 41, it's a safety element out of context (SEooC) in the terminology used in ISO 26262.

** A monolithic operating system is one that incorporates subsystems, such as the file system, communications stack, and device drivers in the kernel; it can be contrasted with a micro-kernel, which only contains essential services.

into a river or the external temperature is -40C. ADC may need to perform a full hazard and risk analysis to determine the correct design safe state for each condition.

Summary

Defining the design safe state(s) of a system or component is not only required by the standards, it is also a very useful exercise for focusing on the role that it is to play in the real world. Not being able to determine the design safe state when the system is in a particular situation may be an indication that the hazards and risks associated with the device have not been fully explored.

References

1. G. Candea and A. Fox, "Crash-only Software," in *HOTOS'03: Proceedings of the 9th conference on Hot Topics in Operating Systems*, (Berkeley, CA, USA), pp. 12–12, USENIX Association, 2003.
2. G. Candea, S. Kawamoto, Y. Fujiki, A. Fox, and G. Friedman, "Microreboot — A Technique for Cheap Recovery," 2004.

Chapter 10

Replication and Diversification

> *The most certain and effectual check upon errors which arise in the process of computation, is to cause the same computations to be made by separate and independent computers; and this check is rendered still more decisive if they make their computations by different methods.*
>
> Dionysius Lardner, 1834

This chapter contains descriptions of several architectural patterns based on replication and diversification, the distinction being that replication deploys two or more identical copies of a system, whereas diversification incorporates different implementations of the same function.

The purpose of using replication or diversification is to increase either the availability or the reliability of a system — see page 14 for an explanation of this distinction. Generally, requiring two systems to present the same response to a particular stimulus will increase reliability while decreasing availability (because both systems have to be running to provide any response).

History of Replication and Diversification

The 1837 quotation from Babbage on page 14 about diverse software design was not quite the first. The quotation from Dionysius Lardner's 1834 paper (reference [1]) given at the top of this chapter predates it by three years. The basic technique, when used by humans rather than computers, dates back millennia for use in counting houses. The Compact Oxford English Dictionary on my bookshelf gives the following as the first definition of "computer": "One who computes; a calculator, reckoner; *spec.* a person employed to make calculations in an observatory, in surveying, etc."

Replication in the Standards

Replication and diversity are techniques recommended in both IEC 61508 and ISO 26262.

Section C.3.5 of part 7 of IEC 61508 describes a technique very close to N-version programming (page 139 below), and the concept of a diversely-implemented external monitor (page 147 below) is listed as a recommended technique in Table A.2 of IEC 61508-3 and described in more detail in section C.3.4 of IEC 61508-7. In edition 1 of IEC 61508 and in EN 50128, this architecture is termed a "safety bag"; in the second edition of IEC 61508, its name has changed to "diverse monitor" which has at least the advantage of being somewhat more descriptive. These are both recommended as "software architecture designs" by IEC 61508.

ISO 26262-6 also recommends the diverse monitor, there called an "external monitoring facility" in Table 4.

The diverse monitor technique is also recommended for medical device software in Table B.2 of IEC/TR 80002-1: *Medical device software — Part 1: Guidance on the application of ISO 14971 to medical device software.* ISO 14971 is included by reference in IEC 62304.

Component or System Replication?

If replication or diversification is to be used, is it better to replicate components or systems: Which of the two designs in the lower part of Figure 10.1 is likely to deliver higher dependability? The answer may vary from system to system, particularly if a system has several different failure modes, but in general, replicating components is better than replicating systems. This result follows from the inequality:

$$\phi(\vec{x} \coprod \vec{y}) \;\geq\; \phi(\vec{x}) \coprod \phi(\vec{y}) \tag{10.1}$$

where the symbols are as defined in Appendix C on page 327.

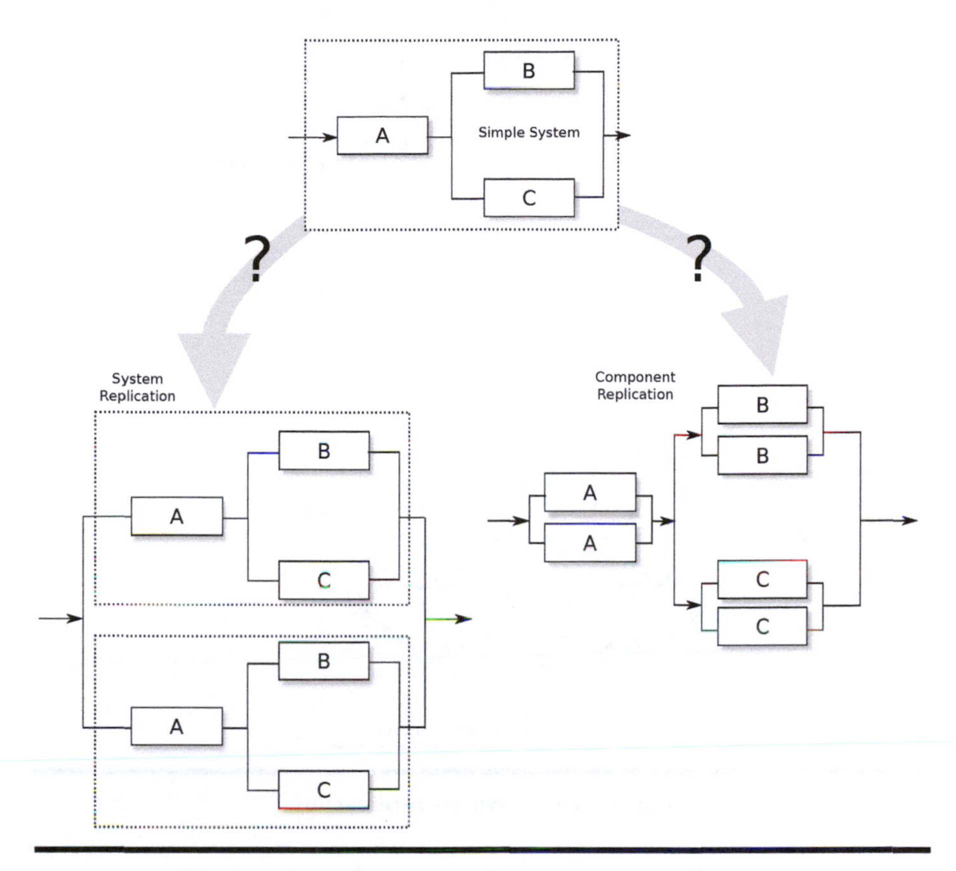

Figure 10.1 System and component replication.

As an extra advantage, if the components are hardware, it may be unnecessary to replicate *all* of them to achieve the necessary dependability, leading to a product that is cheaper than would result from complete system replication.

Replication

System Replication

The drive for replication often stems from the view that, if a single system cannot meet the necessary dependability requirements, then by replicating it, a much higher level of dependability can be obtained. This can be true if the replication is treated very carefully, but often the increased complexity of replication outweighs the advantages.

Consider Figure 10.2. The upper part of that figure shows a simple

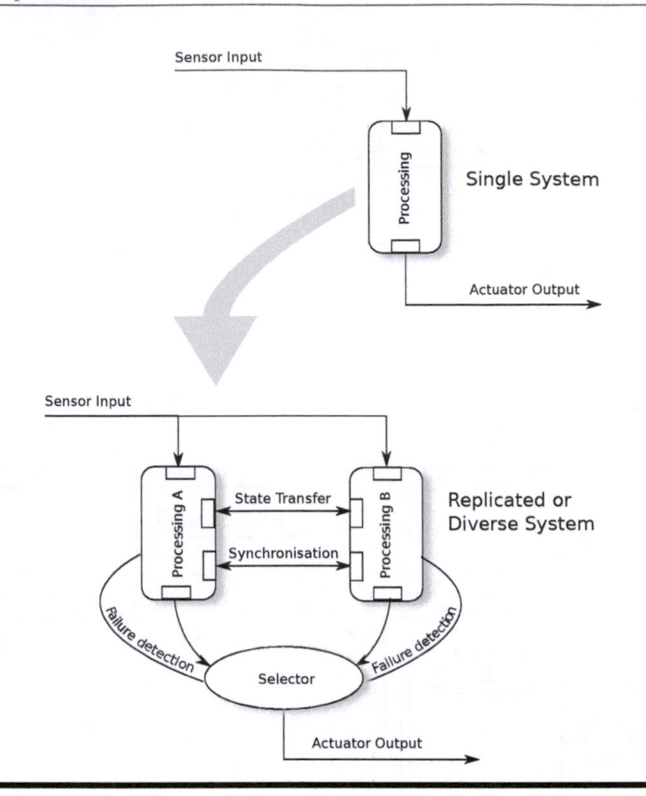

Figure 10.2 Naïve replication.

system that, as most embedded systems, accepts values from sensors, performs some form of computation, and presents some outputs to actuators (displays, relays, lights, etc.). When a failure analysis is performed on this design, it may be found that its dependability is inadequate and a "simple" remedy lies in replication, as shown in the lower part of the figure. As can be seen from Figure 10.2, replication brings additional complexity:

A selector has been introduced to compare the two outputs and choose between them.
If the intention is to increase the system's reliability, then this selector would compare the two outputs and provide output to the actuators only if there is agreement; this reduces system availability, because both subsystems have to be working to allow the comparison to be made.

To increase availability the selector would pass the first input it receives to the actuators; providing no more reliability, but more availability, than the unreplicated solution.

Some form of state transfer may be needed.

If the system stores state, then state will at least have to be shared when a replica starts up — it will need to be brought in line with the operating replica. If a cold standby approach is used (only one of the two subsystems actually performing the calculations), then state transfer may be needed whenever the state changes on the active subsystem to ensure that the standby subsystem has the necessary history (state) to take over at any time.

Some form of synchronisation may be needed.

If the systems are running independently of each other, they may need to co-ordinate responses before presenting them to the selector.

The strength of replication is that it allows the design to be focused cleanly on either availability or reliability. For maximum *availability*, the two (or more) subsystems would both be active and the selector would take the first output provided, a so-called 1oo2 (one out of two) system. For maximum *reliability*, the selector would wait for output from both subsystems and compare them. It would accept the outputs only if they agreed: 2oo2 (two out of two). In 2oo2 mode, if one subsystem failed, then the whole system would stop, availability having been sacrificed for reliability.

One particularly useful form of replication is "virtual synchrony," which is described on page 141 below.

Cold standby

The "cold standby" model is a particular case of the lower part of Figure 10.2, where only one of the subsystems is operating and the other is simply monitoring its behavior, ready to take over if the active partner fails. In this case the selector becomes a switch connecting the active partner to the actuators.

When cold standby is used, some additional potential failure points are introduced:

How is a failure of the idle subsystem detected?

Following the failure of the active subsystem, a switch might be made to the standby only to find that it failed some hours previously. The state transfer module may still be accepting incoming state updates, but the remainder of the software may be unavailable. In many systems, such silent failure is the largest component in the overall system failure probability.

How does the selector (a switch) know that the active subsystem has failed and that a switch to the standby is required?
Does it sometimes switch unnecessarily, and, if so, how does the probability of the cold subsystem having failed silently affect the overall system failure probability? Does it sometimes not detect failure and remain switched to a failed subsystem? Not detecting failure is particularly pernicious, as it means that the overall system has failed while a working subsystem is available and not being used.

How does a failure of the selector itself affect the dependability of the overall system?
Depending on the actual system design, the selector can be a very complex component, itself subject to failure.

Another disadvantage of the cold standby architecture is that it is difficult to scale. If the required level of dependability needs three subsystems, then the selector becomes very complex, and therefore error-prone.

Time Replication

Given that most bugs encountered in the field are Heisenbugs, simply running each computation twice and comparing the answers may be sufficient for the system to detect a bug. Performing the computation a third time and finding two identical answers may then be enough to obtain a reliable result, the adequacy of the design being assessed during the failure analysis — see Chapter 12.

This form of replication is known as "time replication," as the commodity being replicated is time: The system runs more slowly, but more reliably.

Diversification

Diversification takes replication one step further, with the two subsystems in Figure 10.2 somehow different from each other. Possible points of diversification include hardware, software, and data. These are described in the sections below.

Hardware Diversity

It would be possible for the two subsystems in Figure 10.2 to be running on different processors (perhaps an ARM and an x86 processor). The benefit of this is unclear. Certainly, it would entail more work (different board support packages, different board layouts, different compilers, etc.), and we have to consider the types of processor fault against which we are trying to guard: Heisenbugs or Bohrbugs.

Heisenbugs can arise in processors. Every modern processor comes with a book of *errata* that can make the reader's hair stand on end! At the time of writing, the *errata* for one popular processing chip contain statements from the chip manufacturer, such as, "Under very rare circumstances, Automatic Data prefetcher can lead to deadlock or data corruption. No fix scheduled," and, "An imprecise external abort, received while the processor enters WFI, may cause a processor deadlock. No fix scheduled." Even without the chip *errata*, Heisenbugs can arise from electromagnetic interference (see reference [2]), the impact of secondary particles from cosmic ray events or, for many memory chips, "row hammering" (see reference [3]).

However, we do not need diversity to protect against Heisenbugs, we can use replication. If such a Heisenbug affects the operation of one processor, it is extremely unlikely to affect the other processor in the same way at the same time — that's the nature of a Heisenbug.

Diversification of hardware *would* be effective against Bohrbugs in the processor. However, these are very rare. The last reported Bohrbug might be the 1994 division bug in the x86 Pentium processor that could not calculate $\frac{4195835}{3145727}$. To accept the cost of a diverse processor design in case a chip manufacturer has introduced a Bohrbug may not be a cost-effective strategy.

Software Diversity

Simple Code-Level Diversity

There are many levels of software diversity. Perhaps the simplest is to use a tool to transform one source code into an equivalent one. For example, using de Morgan's Laws, the code

```
unsigned char doit(unsigned char x, unsigned char y)
    {
    unsigned char  s;
    s = x & y;
    return s;
    }
```

can be converted into

```
unsigned char doit(unsigned char x, unsigned char y)
    {
    unsigned char  s;
    unsigned char  x1 = ~x;
    unsigned char  y1 = ~y;
    s = x1 | y1;
    return ~s;
    }
```

which is logically equivalent but which, with most compilers, will produce different object code. If both of these representations are executed, the results can be compared before any branching action is taken on the result.

Coded Processors

The idea behind "coded processors" is an extension of the concept of program diversity, whereby a code is stored with each variable and is used to detect computational errors caused by incorrect compilers, hardware errors (bit flips), and incorrect task cycling.

As variables are manipulated (e.g., $z = x + y$ or $x > 0$), so are the appropriate codes. Software or hardware can then check the computation by checking the validity of the new code. This technique is used in the driverless railway system in Lyon, on the Météor line in Paris, and in the Canarsie subway line in New York.

The technique is addressed in IEC 61508-7, paragraph A.3.4, and ISO 26262-5, paragraph D.2.3.6.

Outside the standards, coded processors are described in various publications including reference [4] by Jürgen Mottok *et al.* and reference [5] by Daniel Dollé. The basic idea was developed by P. Forin in 1989 and includes storing a coded form (x_c) of each arithmetical operand (x_f), calculated as:

$$x_c = A \times x_f + B_x + D \tag{10.2}$$

where A is a large prime number, $B_x < A$ is some unique and static signature of x_f (e.g., a value derived from its memory address) that ensures that $z = x + y$ cannot silently be executed as $z = x + x$, and D is a sequence number that increments for each cycle of the computation. The inclusion of D ensures that if one of the variables had been calculated earlier, but its storage in memory didn't happen correctly, perhaps because of a cacheing error, the stale value cannot be accessed without the error being detected.

As two variables, say x_f and y_f, are combined to form z_f (e.g.,

$z_f = x_f + y_f$), the associated coded values x_c and y_c are combined to give z_c by calculating $z_c = x_c \oplus y_c$. The $\oplus$ is the equivalent operation for codes as + is for values. The result of the calculation can then be certified to within $\frac{1}{A}$ (which explains why A should be large) by checking that $z_c = A \times z_f + B_x + D$ or, more simply, that

$$(z_c - B_z - D) \mod A = 0 \tag{10.3}$$

Of course, the extra computation of z_c adds a large overhead to the computation and the storage of a coded value with each variable requires more memory.

Reference [6] by Ute Wappler and Christof Fetzer describes an experiment performed by injecting faults (see page 280) into a program calculating an MD5 hash. Not only did the coded processor technique detect all injected faults that caused the output to be incorrect, it also detected all but 0.06% of the injected faults that resulted in a correct answer, but which corrupted internal state — implying that the next computation might be incorrect.

N-Version Programming

Software diversification can be taken a lot further. N-version programming was proposed by Algirdas Avižienis in 1975 in reference [7]. He defined this technique as follows:

> *N-version programming is defined as the independent generation of $N \geq 2$ functionally equivalent programs from the same initial specification. The N programs possess all the necessary attributes for concurrent execution, during which comparison vectors ("c-vectors") are generated by the programs at certain points. ...*
>
> *"Independent generation of programs" here means that the programming efforts are carried out by N individuals or groups that do not interact with respect to the programming process. Wherever possible, different algorithms and programming languages (or translators) are used in each effort.*

Avižienis countered the argument that N-version programming is prohibitively expensive by saying that the reduced cost of verification and validation can offset the increased cost of programming.

Unfortunately, several studies, including that described in reference [8] by J. C. Knight and N. G. Leveson, have found that code faults are not independent in N-Version programming:

> *For the particular problem that was programmed for this experiment, we conclude that the assumption of independence of errors that is fundamental to some analyses of N-Version programming does not hold. Using a probabilistic model based on independence, our results indicate that the model has to be rejected at the 99% confidence level.*

This is probably because many faults are introduced due to ambiguity in the specifications that are common to all N teams. Moreover most faults appear in complex code and code that is complex for one team is likely to be complex for the others.

The arguments for and against N-version programming caused a serious flame war in the technical literature with passions running very high.

Recovery Blocks

Recovery blocks, as described on page 117, can also be considered a form of diversification.

Data Diversity

Of all the forms of diversity, data diversity is perhaps the most useful, and least controversial.

Anecdote 11 *I once encountered a Bohrbug in an embedded program that occurred when a loop terminated at a value of 1023. In the context, this value was very unlikely and so no failure occurred during testing, but one did occur in the field, with interesting (but, luckily, not dangerous) results. Had the same information been encoded in two different data formats, then the anomaly would have been detected and, possibly, handled.*

The idea is to store the same information using two or more different data representations. In an algorithm for calculating the braking of a train, for example, the same information might be held as data in both the time and frequency domains. This provides a defense against both Bohrbugs and Heisenbugs in programs. A program may, for example, include a Bohrbug that triggers whenever integer overflow occurs. Such overflow might occur while processing the time series, but not when processing the frequency spectrum information.

Virtual Synchrony

Virtual synchrony, also called group synchrony, is a form of replication (or diversification) that can be very powerful in certain architectures. The technique was introduced in the early-1990s by Kenneth Birman and Robbert van Renesse and is described in reference [9].

There are many extensions to the underlying idea of virtual synchrony, but, at its heart, it's based on two very simple concepts:

1. Servers join groups and, when they join, they are provided with an up-to-date copy of the server group's state. Each server may, or may not, be aware of the other members of the group, but each client* making use of the service communicates with the group and believes that it is interfacing with a single server.
2. As client messages arrive and group membership changes (servers joining or leaving), notification of these events is fed to all servers in the group in the same order. So if one member of the group sees that server X has left the group and then sees client Y's request, then all members of the group will see those two events in that order.

The underlying idea is that, if the servers all start in the same state (and the updating of a new group member's state is also part of the protocol) and if each server has received the same messages in exactly the same order, then all the servers must eventually arrive at the same state — this is *virtual* synchrony, not locked-step operation. Most of the algorithmic complexity of virtual synchrony goes into maintaining common views of the current group membership under conditions of server and communications failures.

For some systems, the requirement for common arrival order of all events at each server in the group can be relaxed and different types of guaranteed can be provided:

* I use the term "client" loosely here — it could be the sensors in Figure 10.2 or some other subsystem requesting a service.

Sender order.
All messages from each client are delivered to each group member in the order they were sent. There is no guarantee that messages from two different clients will be delivered in the same order to group members.

Causal order.
If a message s_1 causes a program to send message s_2 (i.e., s_1 causes s_2), then s_1 will be delivered before s_2 to all group members.

Total or agreed order.
If any group member receives s_1 before s_2, then all group members will. This is the ordering described above and it is the ordering assumed in the examples below.

As an example, consider Figure 10.3. In this system, there are four server instances (scaling in the number of servers is not a problem). These could be replicas or could be diversely implemented although there may be little reason to include diversity.

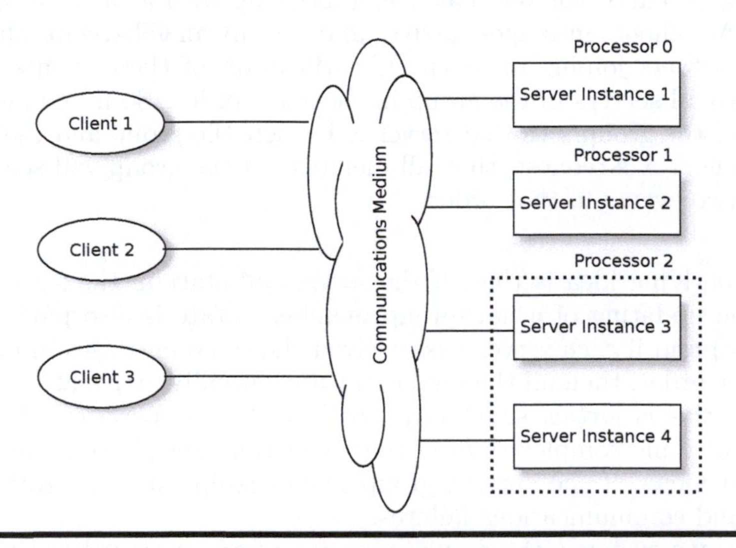

Figure 10.3 Example virtual synchrony environment.

As indicated in Figure 10.3, the designer has decided to run servers 3 and 4 on the same processor.* This is to ensure that the server

* The term "processor" is used to indicate any processing element: it might be a processor chip or a core on a multicore processor.

instances do not run synchronously. Actually, given the attention to detail that is required to force locked-step operation, this is probably not necessary — servers on different processors will diverge after a little time in any case.

I have indicated the communication mechanism between the clients and the servers in Figure 10.3 by a cloud. In principle, this can be any medium supporting messages: a wide-area network (WAN), a local-area network (LAN), a backplane in a rack, tracks between processors on a printed-circuit board, or internal transfer mechanisms between cores on a multicore processor. The closer the connection, the better virtual synchrony works.

Anecdote 12 *I was once called in to consult about a system operating virtual synchrony over a WAN. Because of the long and highly variable delays across the WAN and the high probability of packets being lost in transit, the virtual synchrony algorithms were spending a disproportionate amount of time recalculating and redistributing group membership information. In the end, the project was cancelled.*

There are three clients in Figure 10.3. Say that client 1 and client 3 send in a request to what they believe is a single server, at more or less the same time. The virtual synchrony middleware, of which more later, replicates and orders the incoming requests and ensures that each of the four server instances receives the requests in the same order.

Each server believes that it is alone and so performs whatever computation is required and replies with an answer. The virtual synchrony middleware accepts these answers, which arrive at different times, and takes a predefined action. In general, the action would be to send the first answer back to the client (or on to actuators) and discard the others. However, it could be configured to wait for a majority of server responses, compare them, and send the agreed value to the client.

In any case, the client has submitted a single message to what it believes is a single server and has received back a single response.

This technique has certain advantages over other replication schemes:

It is tolerant to Heisenbugs.

This may be its primary advantage. Assume that, in Figure 10.3, server instance 2 hits a Heisenbug while handling client 3's request and crashes. It is unlikely that the other instances will hit the same Heisenbug: That's why it's called a Heisenbug! In this case, client 3 will disappear from the group, and a response from one of the other server instances will be returned to the client. This is handled in the virtual synchrony middleware without either the clients or the remaining servers being aware.

It scales well (linearly) in the number of server instances.
This allows any level of dependability to be provided. In contrast, as reference [10] by Jim Gray *et al.* points out, the time required to handle a transaction increases as N^3 in other replication architectures.

It can be tuned to provide reliability or availability.
By changing the algorithm of how the middleware handles responses, the system can be configured to provide availability or reliability. For example, discarding all but the first response increases availability at the cost of reliability; waiting for the last response and comparing them all increases reliability at the cost of availability. Selecting algorithms in between (e.g., waiting for the first two responses) allows some control over the availability/reliability balance.

Silent failure is avoided.
All instances of the server are active all the time, reducing the risk of silent failure that bedevils the cold-standby architecture.

The selector or switch in Figure 10.2 is removed.
As all server instances are active at all times, there is no switch that could provide another point of failure.

Generally, no significant change is needed to either the client or the server to introduce virtual synchrony.
The technique can be introduced easily to an existing, non replicated system that is not providing the necessary level of dependability because the client software can continue to believe that it is accessing a single server, and each server can be unaware of the other group members.

It is a standardized technique.
While not referred to explicitly in ISO 26262 or IEC 61508 (being covered by the general heading of "replication"), it is treated in the Service Availability Forum's *application interface specification* (AIS), available from `http://www.saforum.org`.

State transfer is avoided.
Except when a new member joins a group, there is no need to copy state from an "active" to a "passive" server as there is with a cold-standby architecture.

Of course, virtual synchrony also has its disadvantages. First, all state changes in the server must take place as a result of received messages to ensure that the server states do not diverge. This means, for example, that a server instance may not generate a random number locally if that could alter its final state.

Secondly, it wastes processor time — in Figure 10.3, four servers are all performing the same calculation! There are mechanisms within virtual synchrony to reduce this level of work, particularly if no server state change results or if the computation is long and the state change is small. But, in general, work will be replicated.

The third disadvantage would superficially seem to be the most important: The whole concept of group membership on which virtual synchrony is based, is impossible — see reference [11] by Tushar Chandra *et al.* Consider Figure 10.3. If server instance 1 loses contact with the other instances, it is impossible for those other instances to determine whether this is because instance 1 has crashed or because, although instance 1 is still running, communications have been broken to it.

If the communications channels are long and fragile, as in Anecdote 12 above, then this is a really significant problem. Instance 1 may still be alive and providing service to some of the clients, while other clients are connected to the group containing instances 2 to 4. In this way, the system has split and the two parts are diverging. Such communications failures are less likely as the channels get shorter — to a LAN, a backplane, a track on a circuit board, etc. — and the logical impossibility of coherent group membership is no longer so important.

Being based on an impossibility would seem to be a significant disadvantage, but there are pragmatic ways of avoiding the problem. One pragmatic, but imperfect, way of handling the impossibility is for the smaller subset of server instances (instance 1, in this case) to kill themselves if they lose contact with more than half the group. In the case of the example, instance 1 would go from being 1 of a group of 4 to being 1 of 1 and so would deliberately fail. Instance 2 would go from being 1 of a group of 4 to 1 of a group of 3 and would remain running.

In summary, the virtual synchrony pattern is a tested, proven, and standardized way of providing replication in such a way that the system is resilient to Heisenbugs. It is particularly useful when dependability has to be "added" to enhance an existing system.

There are several commercial and open-source implementations of the middleware, including *Spread* (see `http://www.spread.org`), which is specifically designed to operate over WANs, and openAIS (see `http://freecode.com/projects/openais`) which implements the Service Available Forum's AIS.

Anecdote 13 *I first worked on a virtual synchrony system in 1994. This was not an embedded device; rather, the nodes consisted of workstations connected by a LAN. At one point, the cable carrying the LAN went under a door and, inevitably, the cable finally broke at that point. The system had been running without a system-level failure for some years when this occurred, but, unfortunately, we had not been fully aware of the need to configure for this problem, and the LAN was split exactly into two — so both halves shut down as described above. To prevent this from happening again, we redistributed the logical weights associated with the nodes so that the nodes could not be split exactly in half again (and rerouted the cable).*

The algorithms used to maintain a coherent view of the group membership across all members of the group as members fail and recover are very simple, but very subtle. Many published algorithms have been found to contain flaws that could affect system operation in unlikely circumstances. Reference [12] by Massimo Franceschetti and Jehoshua Bruck contains a very simple, and apparently unchallenged algorithm for maintaining views of group membership. Reference [12] recognizes the logical impossibility of what it is trying to do (reference [11]), and minimizes, while obviously not eliminating, the effects of this.

Locked-Step Processors

One surprisingly popular form of replication has two processors running the same code in locked step, each executing the same machine code instruction at the same time.

Anecdote 14 *In the 1980s, I worked on a system in Switzerland that had duplicated processors and triplicated memory. The processors ran in locked step, and the memory was further protected by 8 parity bits for each 32-bit word. The computer consisted of hundreds of wire-wrapped boards; it is the only computer I have ever programmed that required a ladder to maintain.*

Running on locked-step processors has had an interesting history. In the 1980s, software was very simple, and processor hardware very complex and prone to error — in the case of the system mentioned in Anecdote 14, the locked step provided a useful way of detecting a hardware fault, in particular the most common type of bug in those days, a connector failure.

In the late 1990s and early 21st century, hardware became much more integrated, and the number of soldered and wire-wrapped connections reduced significantly. To a large extent, the processor hardware could be removed from the failure analysis because its lifetime was now measured in hundreds of years. At this time, locked-step processing was useless because almost all faults were in software and, with locked step processors, both processors would fail simultaneously for both Heisenbugs or Bohrbugs.

In the last few years, the track widths in the integrated circuits have reduced. In 2005, track widths of 100 nm were typical; by 2010, this had reduced to 32 nm, and by 2018, tracks of 8 nm are predicted. Because of this reduction, processors and memory chips have become more susceptible to cross-talk between tracks, thermal aging, electromagnetic interference, and cosmic ray effects. Locked step processing may again have a positive effect on the reliability aspect of the failure analysis, while having a negative effect on the availability. It still will not, of course, catch either form of software bug.

Diverse Monitor

The External Monitor Pattern

Figure 10.4 illustrates the normal architecture for an external monitor (sometimes called a "safety bag"). This architecture has been used extensively in railway signaling systems dating back to the 1980s — for example see reference [13] by Peter Klein.

The main system performs its function of reading input values and transforming these into output values. The monitor observes the inputs and outputs, and may also observe internal states of the main system.

The difference between the main system and the monitor is that the main system is *much* more complex. For example, in a railway braking application, the main system may accept speed and distance inputs from sensors and movement limits ("you are authorized to travel another 34 km from this position") from track-side equipment. Taking into account the gradient and curves of the track ahead, the current speed, and dozens of other factors, it calculates the optimal point at

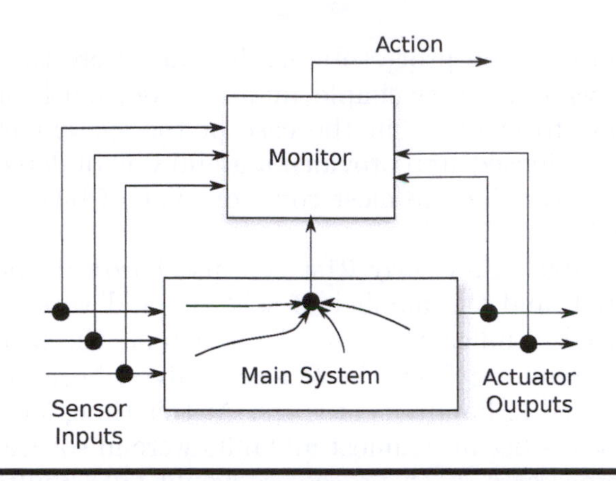

Figure 10.4 Diverse monitor.

which braking should occur if no further movement limit is received. It notifies the driver of this using a sophisticated graphical system and, if the driver does not respond appropriately, takes over control and applies the brakes. There are many places for potential failure in this chain.

The monitor, in contrast, is *very* simple. In many systems, it may be implemented in hardware without any software, possibly using a field-programmable gate array (FPGA). In the train braking example, it could also read the movement limits and speed, but its calculations are much simpler. For example, if the train is within two minutes of its movement limit, the monitor applies the brakes. If it can be demonstrated that, whatever the main system does, the monitor keeps the train safe, then it may be necessary to certify only the monitor — this becomes the safety element.

When designing an automobile system, the automotive safety integrity level (ASIL) decomposition provided by ISO 26262-9 may often be applied. For example, if the complete automotive system is to be certified to ASIL C, then, subject to certain conditions, paragraph 5.4.10 of ISO 26262-9 permits this to be decomposed into an ASIL C component and a "quality management" (QM) component to which ISO 26262 does not apply. In Figure 10.4, the main system could perhaps be governed by the QM processes and the monitor by the ASIL C processes.

Figure 10.5 shows another configuration of the monitor. It is now inline with the outputs from the main system and can change them before they reach the actuators.

The monitor is again *much* simpler than the main system, and it applies a sanity check on the main system's outputs. In a medical

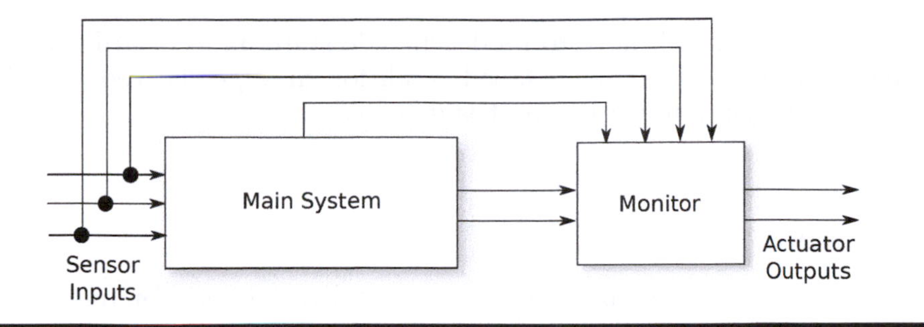

Figure 10.5 Diverse monitor: alternative configuration.

device, for example, the main system may be calculating and applying a drug flow. While the monitor is less sophisticated than this, it may be able to detect that, given the input conditions, the drug flow configured by the main system cannot possibly be valid and can cap it to a safe value.

In general, whichever of the configurations are used, the monitor is providing non optimal, but nevertheless safe, behavior.

Watchdog Diverse Monitor

One very common diverse monitor is the watchdog. In the configuration shown in Figure 10.4, the main system is required to "kick" the watchdog (the monitor) periodically by sending it a message or by changing a memory value. If the watchdog notices that it hasn't been kicked for a pre-determined time, then it "barks," typically forcing the system into its design safe state.

The difficulties with the watchdog approach are finding a suitable place within the main system's code to kick the watchdog and setting the time within which the watchdog must be kicked.

Clearly, whatever is kicking the watchdog must stop doing so if a failure has occurred in the main system, but it is tempting to put the watchdog kicker in a timer interrupt routine. This provides almost no protection because, even if almost all the code in the main system has failed, the timer interrupt may still be arriving and the watchdog may still be being kicked. To avoid this problem, the process kicking the watchdog must be sophisticated enough to be able to determine whether the system is working correctly or not and this may not be easy to determine.

There is also a compromise to be made regarding the time after which the watchdog will bark. If this is set too low, then unusual, but correct, behavior on the main system may cause a delay in the kicking, and the system will move to its design safe state unnecessarily. If the time is set too high, then a failed condition may persist too long for safety to be guaranteed before it is detected.

Summary

Replication and diversification are powerful techniques for improving the availability or reliability of a system. However, their use is not free — they add extra components and complexity to the system, and an unthinking application of replication may actually result in a decrease in dependability. Silent failure of a component that is in cold standby mode is particularly pernicious.

References

1. D. Lardner, "Babbage's Calculating Engine," *The Edinburgh Review*, vol. 59, 1834.
2. K. Armstrong, "Including Electromagnetic Interference (EMI) in Functional Safety Risk Asessments," in *Proceedings of the 20th Safety-Critical Systems Symposium*, SSS'12, pp. 97–114, Springer-Verlag, 2012.
3. Y. Kim, R. Daly, J. Kim, C. Fallin, J. Lee, D. Lee, C. Wilkerson, K. Lai, and O. Mutlu, "Flipping bits in memory without accessing them: An experimental study of DRAM disturbance errors," in *ACM/IEEE 41st International Symposium on Computer Architecture, ISCA 2014, Minneapolis, MN, USA, June 14-18, 2014*, pp. 361–372, 2014.
4. J. Mottok, F. Schiller, T. Völkl, and T. Zeitler, "A Concept for a Safe Realization of a State Machine in Embedded Automotive Applications," in *Proceedings of the 26th International Conference on Computer Safety, Reliability, and Security*, SAFECOMP'07, (Berlin, Heidelberg), pp. 283–288, Springer-Verlag, 2007.
5. D. Dollé, "Vital software: Formal method and coded processor," in *3rd European Congress on Embedded Real Time Software, ERTS 2006, Toulouse, France, January 25-27, 2006*, 2006.
6. U. Wappler and C. Fetzer, "Hardware Failure Virtualization Via Software Encoded Processing," in *5th IEEE International Conference on Industrial Informatics (INDIN 2007)*, vol. 2, pp. 977–982, IEEE Computer Society, June 2007.
7. A. Avižienis, "Fault-tolerance and fault-intolerance: Complementary approaches to reliable computing," in *Proceedings of the International Conference on Reliable Software*, (New York, NY, USA), pp. 458–464, ACM, 1975.
8. J. C. Knight and N. G. Leveson, "An Experimental Evaluation of the Assumption of Independence in Multi-Version Programming," *IEEE Transactions on Software Engineering*, vol. SE 12, no. 1, pp. 96–109, 1996.

9. K. P. Birman, *Building Secure and Reliable Network Applications.* Upper Saddle River, NJ, USA: Prentice Hall PTR, 1st ed., 1996.
10. J. Gray, P. Helland, P. O'Neil, and D. Shasha, "The Dangers of Replication and a Solution," in *In Proceedings of the 1996 ACM SIGMOD International Conference on Management of Data*, pp. 173–182, 1996.
11. T. D. Chandra, V. Hadzilacos, S. Toueg, and B. Charron-bost, "On the Impossibility of Group Membership," 1996.
12. M. Franceschetti and J. Bruck, "A Group Membership Algorithm with a Practical Specification," *IEEE Trans. Parallel Distrib. Syst.*, vol. 12, pp. 1190–1200, Nov. 2001.
13. P. Klein, "The Safety-Bag Expert System in the Electronic Railway Interlocking System ELEKTRA," *Expert Systems with Applications*, vol. 3, pp. 499–506, 1991.

DESIGN VALIDATION

Chapter 11

Markov Models

If there is a 50-50 chance that something can go wrong, then 9 times out of 10 it will.

Paul Harvey

There is one implicit assumption underlying all failure models of software systems — the assumption that we can predict with some degree of confidence the failure rate of the software components. Certainly, to produce any form of failure analysis of a system containing software, including the Markov analysis in this chapter and the fault tree analysis in Chapter 12, it is essential to be able to predict software failure rates. Preparing such predictions is an important topic and is covered in Chapter 13.

Markov Models

Markov chains were introduced in Chapter 8 starting on page 102 as a means of detecting anomalies in system operation. This chapter deals with their use for estimating the failure rate of a system. Both of these techniques are based on the same concept: finding the steady-state probabilities of the system being in particular states as it moves between states according to probabilities associated with the different transitions.

The models (chains) to be considered in this chapter are much simpler than those likely to be encountered during anomaly detection. Generally, they include sufficiently few states that transition diagrams such as Figure 11.2 can be drawn.

Markov Models and the Standards

Markov models are mentioned in part 9 of ISO 26262, along with fault tree analysis, as a quantitative tool suitable for conducting a safety analysis. Markov models are also explicitly invoked in Table 1 of part 4 (product development at the system level) of ISO 26262 as a suitable tool for inductive analysis to avoid systematic failures.

IEC 61508 is much more enthusiastic about Markov modeling. Section B.6.6.6 of part 7 describes the technique and provides several references. However, the authors of this section of IEC 61508 are aware of the limitations imposed by the Markovian assumptions (see next section) and say, "When non-exponential laws have to be handled — semi-Markov models — then Monte-Carlo simulation should be used." Monte-Carlo simulation is the subject of the section on discrete event simulation in this book, starting on page 190.

The Markovian Assumptions

Dependability analysis by Markov modeling is extremely straightforward, but the price paid for its simplicity is that it makes assumptions; the analyst needs to assess the applicability of these before applying the technique to a particular system. I believe that it is a useful tool for carrying out the quick assessment of whether a design could possibly be adequate, as shown in the *initial design cycle* of Figure 1.1 on page 8.

Discrete event simulation (page 190) and Petri nets (page 199) make fewer assumptions but are harder to apply.

In this chapter, the Markovian technique is described through an example. For any Markovian analysis to be valid, the following assumptions must be made about the system under analysis:

When the system is in a particular state, the history of how it arrived in that state is irrelevant for determining what will happen next. This means that this method of estimating is only valid when the failure rate is assumed to be constant.

For example, take a system with three components: A, B, and C. If the current system state is that A is working and B and C have failed, the future of the system does not depend in any way on whether B or C failed first. If there are such dependencies then the number of states needs to be increased:

- A working, B and C failed, with B having failed first
- A working, B and C failed, with C having failed first

This type of subdivision can rapidly cause an explosion in the number of states in the model.

The interarrival times of failure events must be negatively exponentially distributed.

This is actually a direct corollary of assumption 11, because the negatively exponential distribution is the only memoryless continuous distribution. Such a process is known as a Poisson Process.

Its implications can be somewhat nonintuitive. For example, consider a hardware system consisting of components A and B, each of which runs continuously and each of which has a mean time to failure of 1 year. The system has been running for 11 months and component A has just failed (not unreasonably given its mean time to failure). How much life does B have left? For a memoryless system, the answer has to be 1 year.

In spite of this apparent flaw, failures occurring with negatively-exponentially distributed interarrival times have been used to model mechanical hardware failure, although they may be considered more appropriate to software, because it doesn't wear out. If components A and B in the example above were software modules, we would not blink at B's expected life still being 1 year — it would not have worn out. In reference [1], Norman Fenton and Martin Neil point this out (section 12.3) and note that the very fact that software doesn't wear out makes a Poisson Process a useful model for representing its failure.

Even if software failures cannot be modeled completely as a Poisson Process, all may not be lost because, by introducing additional states, other types of distribution can be produced. For example, rather than specifying an exponentially distributed mean time between failures of 5 years, it may be better to specify the mean time to failure as the sum of two exponentially distributed mean times between failures of 2.5 years. Unfortunately, this again leads to an increase in the number of system states.

Only one state change may occur at any time.

This is not normally a problem because the time unit can be made arbitrarily small.

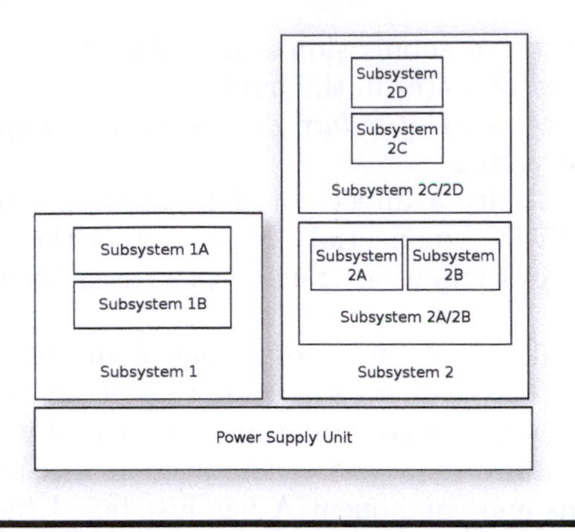

Figure 11.1 A simple system.

Example Calculation

Estimating the Rates

Figure 11.1 illustrates a simple system with 7 components that may fail: 1A, 1B, 2A, 2B, 2C, 2D, and the power supply. In order to continue operating, the system needs the following to be working:

- the power supply, **and**
- subsystem 1, which will only work if both subsystems 1A **and** 1B are operational, **and**
- subsystem 2, which will only operate if 2C **and** 2D are operational **and either** 2A **or** 2B (or both) is operational.

We will take our time unit to be a year and will take the mean failure rate of all the components except the power supply to be 0.333 per year (i.e., a failure on average every 3 years). The mean failure rate of the power supply is 0.2 per year.

The other values that we will need are repair times: When a component failure occurs, how long does it take for a repair to be completed? We will be making the unrealistic assumption that all failures, even those that do not immediately bring the entire system down, will be detected.*

* It is failing to detect component failures that normally contributes most to the failure rate of the complete system.

For the purposes of this exercise, we will assume that the mean repair time for a failed component when the system is still running is 2 weeks (i.e., a rate of 26 per year), and the mean repair time for a component when the system has failed is 1 day (i.e., 365.25 per year). When a repair technician arrives on site, all the failed components are fixed.

Note that all the failure and repair rates are means; the actual times to failure or repair will be negatively exponentially distributed with those means.

This example is analyzed further in Chapter 12 where a fault tree is drawn for it.

Identifying System States

Figure 11.2 illustrates the states between which the system may move.

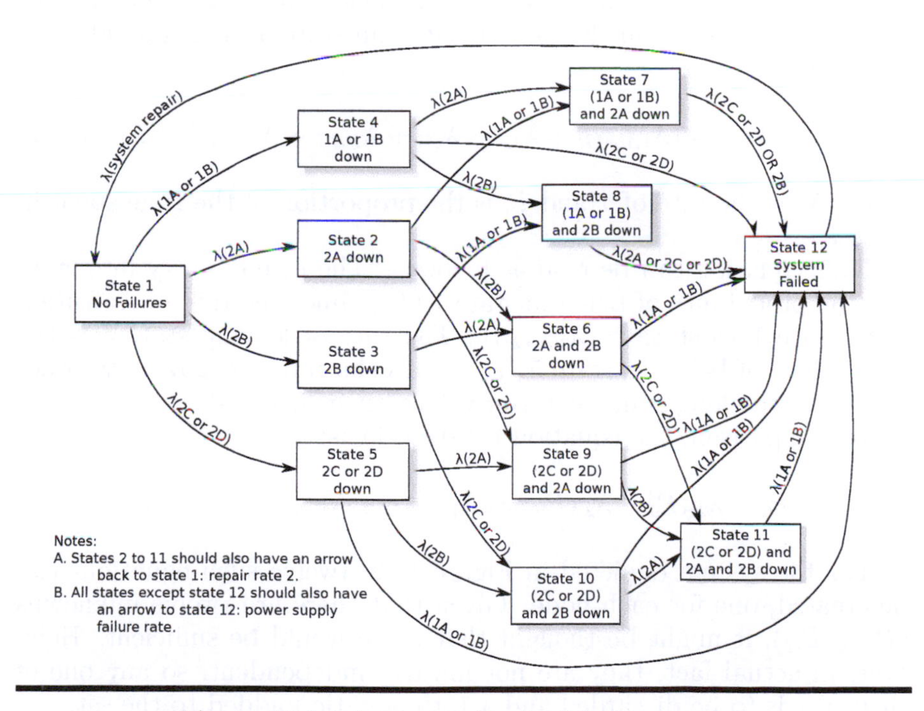

Figure 11.2 An example of a Markov model.

Note that, as indicated in the figure, some arrows have been omitted to reduce clutter.

Given those omissions, Figure 11.2 can be read as follows. Assume that the system is in state 5 with component 2C or 2D failed. From this state it can move to state 12 (system failed) if either 1A or 1B fails; it can move to state 9 if 2A fails; it can move to state 10 if 2B fails; and it can move to state 1 if the repair is completed (arrow not shown).

In accordance with assumption 11 above, only one state change may occur at a time. It is therefore unnecessary to model, for example, the condition that, when in State 1, components 1A and 2A fail simultaneously. Only one event can occur at each moment in time.

Note that a Petri net provides an alternative way of expressing Figure 11.2 — see page 205.

Building the Equations

Given the rates at which each of these transitions occur, a series of simultaneous linear equations can be created to reflect these state transitions. Given state 5 as an example, over the long term, the rate of entry into state 5 must equal the rate of exit from state 5. This provides the equation:

$$\lambda_{2Cor2D} \times P_1 = (\lambda_{1Aor1B} + \lambda_{2A} + \lambda_{2B} + \lambda_{PSU} + \lambda_{repair}) \times P_5 \quad (11.1)$$

where λ_x is the rate of x and P_i is the proportion of the time spent in state i.

This equation can be read as follows: "The rate of entry into state 5 is the probability of being in state 1 (P_1) times the rate of transition from state 1 to state 5 (λ_{2Cor2D}). The rate of leaving state 5 is the probability of being in state 5 (P_5) times the rates of moving to other states. In the long run, these two values must be equal."

In comparison, the equation for state 11 is:

$$\lambda_{2Cor2D}P_6 + \lambda_{2B}P_9 + \lambda_{2A}P_{10} = (\lambda_{1Aor1B} + \lambda_{PSU} + \lambda_{repair})P_{11} \quad (11.2)$$

For the system depicted in Figure 11.2, twelve such equations can be created, one for each node. Given that there are twelve unknowns ($P_1 \ldots P_{12}$), it might be thought that these would be sufficient. However, in actual fact, they are not linearly independent, so any one of them needs to be discarded and a 13th equation added to the set:

$$\sum_{i=1}^{12} P_i = 1 \quad (11.3)$$

This equation expresses the condition that, at any time, the system will be in exactly one of the twelve states.

Solving the Equations

The resulting twelve simultaneous linear equations can be solved using any convenient method (e.g., LU-decomposition). In selecting a method, note that there can be a wide disparity in the magnitude of numbers that occur in this type of analysis. Even in this trivial problem, we have numbers as big as 365.25 (the repair rate following system failure) and as small as 0.2 (the failure rate of the power supply). In a more complex system, failures may occur only every few years, and repairs (particularly in a software system) may occur within milliseconds. Unless care is taken in the solution of the equations and unless the frequency unit is chosen appropriately, resolution can easily be lost.

In this case, creating and solving the twelve simultaneous equations gives the information listed in Table 11.1.

Table 11.1 Solution of the example Markov model.

State	Percentage Time in State	Visit Frequency (per year)	Mean Visit Duration
State 1	92·963044	2·045	0·45458678
State 2	1·111894	0·309846	0·035885382
State 3	1·111894	0·309846	0·035885382
State 4	2·250707	0·619692	0·036319788
State 5	2·250707	0·619692	0·036319788
State 6	0·026920	0·00741188	0·036319788
State 7	0·054829	0·0149135	0·036764706
State 8	0·054829	0·0149135	0·036764706
State 9	0·054829	0·0149135	0·036764841
State 10	0·054829	0·0149135	0·036764841
State 11	0·002028	0·000544939	0·037220936
State 12	0·063491	0·2319	0·0027378508

Analyzing the Results

There are some important points to be made about these results. Firstly, we can perform some sanity checks on them; for example, the mean visit duration to State 12 is 0.0027378508 years. That resolves to 1 day and is what would be expected, the assumed repair time for a failed system being 1 day. It can also be seen that the average time spent in states 7, 8, 9, and 10 is the same. This is also to be expected because the conditions to enter and exit those states are equivalent.

The second point is much more important. State 12 is our system failure state, and we know that we spend 0.063491% of the time in it. So our operational time is 100.0% − 0.063491% = 99.936509%, isn't it? Of course it isn't. We've taken failures rates estimated to a few decimal places and likely to be significantly in error, and we are quoting our up-time to 8 decimal places! This is a trap into which even the best analysts fall from time to time. In fact, we should perform a sensitivity analysis on the equations and publish only the number of decimal places that can be supported. For this example, the best we can probably say is that the operational time will be about 99.9%. Remember Norman R. Augustine's 35th law: "The weaker the data available upon which to base one's conclusion, the greater the precision which should be quoted in order to give the data authenticity."

The temptation to believe the arithmetic was noted in the *Fault Tree Handbook* (reference [2]) as early as 1981:

> *...then the probabilities of events that we are ignoring become overriding, and we are suffering from the delusion that our system is two orders of magnitude safer or more reliable than it actually is. When due consideration is not devoted to matters such as this, the naive calculator will often produce such absurd numbers as 10^{-16} or 10^{-18}. The low numbers simply say that the system is not going to fail by the ways considered but instead is going to fail at a much higher probability in a way not considered.*

Markovian Advantages and Disadvantages

Markov modeling can be quick and easy to perform and has the advantage of providing precise and exactly reproducible results. Its disadvantage is that it assumes that failure rates are constant, the interarrival times of failures being negatively exponentially distributed.

There is no question that the failures of the system under design *may* occur with that distribution. So if the Markov modeling indicates that the failure rate is too high, then it is. However, if the Markov model produces an acceptable failure rate, then that is not a result that can be trusted for the general case: A Markov model can always produce the evidence for a design to be rejected, but can never cause it to be accepted.

References

1. N. Fenton and M. Neil, *Risk Assessment and Decision Analysis with Bayesian Networks.* CRC Press, 2013.
2. W. Vesely, F. Goldberg, N. Roberts, and D. Haasl, "Fault Tree Handbook," 1981. National Research Council Report NUREG-0492.

Chapter 12

The Fault Tree

Trees're always a relief, after people.

David Mitchell

Although the Markovian technique described in Chapter 11 may be adequate for the *initial design cycle* of Figure 1.1 on page 8, once a candidate design is available, a more accurate failure analysis is needed, and this is created during the *detailed design cycle*. This chapter investigates the use of fault tree analysis (FTA) for that purpose. As well as describing a conventional Boolean fault tree, it also covers the more powerful and flexible technique of Bayesian fault tree analysis.

Any failure analysis of a system needs as input the failure rates of the individual components of the system, and the question arises of how to estimate the failure rate of a software component. This is an important topic and is covered in a chapter of its own — Chapter 13.

Some of the techniques described in Chapter 14, particularly discrete event simulation and Petri nets, are complementary to fault tree analysis and can also be used to assess a system's dependability.

FTA and FMECA

FTA was compared with failure mode, effect, and criticality analysis (FMECA) on page 67. Personally, I have found the FTA approach to be much more useful for expressing the failure model of systems, particularly systems relying on the continued correct operation of software. However, I am aware that some people prefer FMECA, and that is certainly an acceptable alternative.

Fault Tree Analysis in the Standards

IEC 61508-3 recommends the use of FTA in Table B.4, which is invoked from Table A.10.

Both FTA and FMECA are described in Annex B of ISO 26262-10, where a method for combining them is provided, and in Annex E of EN 50126-2.

Annex G.3 of ISO 14971 recommends the FTA technique for medical devices.

As mentioned on page 42, IEC 62304 contains, in section 4.3, one very enigmatic quotation: "If the hazard could arise from a failure of the software system to behave as specified, the probability of such failure shall be assumed to be 100 percent." This requires that the assumption that software will always fail be included in the failure model, but gives no guidance about the frequency or type of failure: Should it be assumed that the software fails every 100 ms, or every week?

Types of Fault Tree

The following three sections illustrate fault trees of increasing sophistication and flexibility.

The first is a classical fault tree, which allows the designer to represent the Boolean (true/false) relationship between component and system failure: "If X *AND* Y both fail, then subsystem Z fails".

This concept is extended in the second example to permit statistical reasoning: "if X *AND* Y both fail, then subsystem Z fails. X has a failure rate of λ_x, and Y has a failure rate of λ_y."

This is still not flexible enough to express two very common situations:

If X or Y fails, then subsystem Z will normally, but not always, fail. This construction is particularly suitable for situations where a human is involved. For example, it might be useful to express the relationship that if an operator enters the wrong drug dosage, then the patient may die (a failure of the system to protect the patient). But it is not inevitable that entering the wrong dosage will hurt the patient — some wrong dosages (e.g., 10% below the correct dosage) may be relatively harmless. It might be convenient to express the condition that entering a wrong dosage causes a system failure in 90% of cases (this number having been obtained through experiments in a hospital).

X and Y may not be the only things that can contribute to the failure of subsystem Z.
Particularly in the early stages of analyzing the failure of a complex system, there may be other factors that have not been included, but it is felt that X and Y probably cover *most* of the factors causing a failure of Z. However, even at the end of a project, it is a bold analyst who claims that every possible factor has been identified.

These conditions are expressed in the third example below by using the so-called *noisy conjunctions.*

Example 1: Boolean Fault Tree

Traditionally, an FTA has led to a Boolean fault tree where each node has a value of either TRUE or FALSE. There are many concepts that cannot be expressed in this simplified manner, and so I prefer to use a Bayesian fault tree. However, let us explore the Boolean fault tree first.

Consider the system introduced on page 158 that has 7 subsystems, called 1A, 1B, 2A, 2B, 2C, 2D, and "power supply." These may be software or hardware subsystems without a change to the fault tree.

Figure 12.1 illustrates a fault tree consisting of 7 nodes corresponding to the subsystems and a number of AND and OR gates. An AND gate is true (representing failure at that point) if all of its inputs are true; an OR gate is true unless all of its inputs are false.

Figure 12.1 shows at the top level that the system fails (true) if both subsystem 1 **and** subsystem 2 fail — presumably due to the incorporation of some kind of redundancy into the design. Subsystem 1 fails if any one of 1A, 1B, **or** the power supply fails. And so on.

In Figure 12.1, I have created a "scenario" for each of the leaf nodes where 1B and 2B have failed (they are true) and the other components are still operating. Note that the system as a whole is still functioning because the failure of these two subsystems is not sufficient to bring down the whole system.

This observation leads us to the concept of *minimum cut sets*: the smallest sets of nodes that, by failing, will bring the system down. In the case of Figure 12.1, the minimum cut sets are: { power supply }, { 1A, 2C }, { 1A, 2D }, { 1B, 2C }, { 1B, 2D }, { 1A, 2A, 2B }, and { 1B, 2A, 2B }.

Knowledge of the cut sets is of great help to the designer because it allows focus to be placed on the more critical elements (here, this is ob-

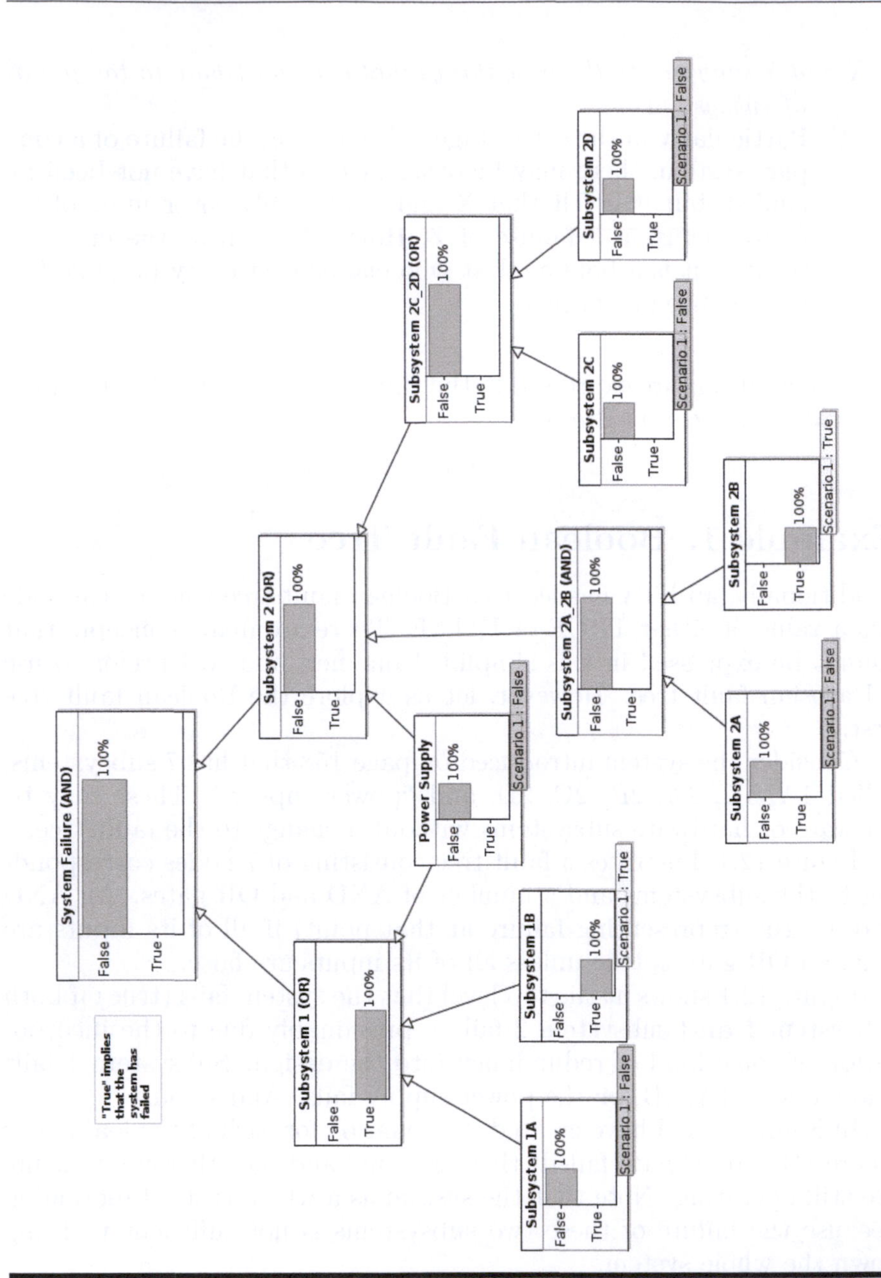

Figure 12.1 A simple Boolean fault tree.

viously the power supply, but in other designs, the critical components might be harder to identify).

The other use of the minimum cut sets is to create a canonical* form of a system design — see Figure 12.2. With such a figure, any combination of failures that causes the left-to-right path to be completely broken leads to a system failure. The canonical form can be used as the basis of the structure function described on page 329.

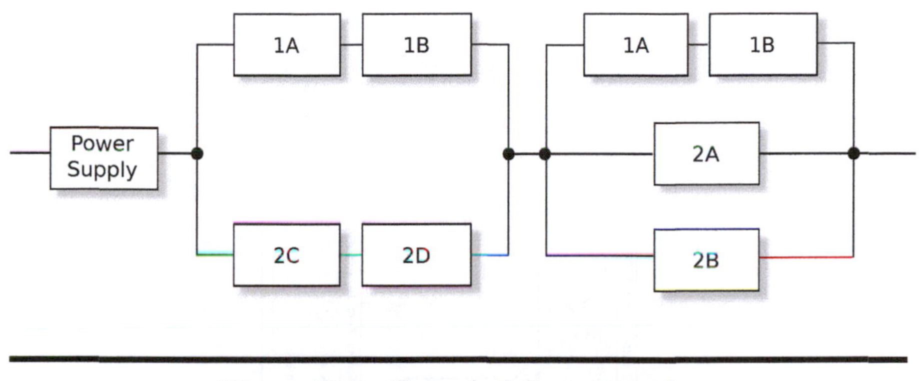

Figure 12.2 Canonical form example.

Example 2: Extended Boolean Fault Tree

The very simple fault tree in Figure 12.1 is, by itself, limited in use. It provides us with the minimum cut sets, which are themselves useful, but little else.

In Figure 12.3, I have extended the Boolean fault tree slightly by including probabilities of failure per N hours of operation. N is often chosen as 10^9 (about 114,077 years), and the failure rate is then measured in "FITs," a failure rate of 68 per 10^9 hours being written as "68 FITs."

In Figure 12.3, I have used a very simple way of encoding the values of each node: How likely is it that the subsystem will have failed in 10^9 hours of operation? For example, the values entered against Subsystem 2A indicate that it has been assessed that the probability of failure in 10^9 hours is 5%. Of course, in real systems, the probability distribution may be much more complex, and most FTA tools allow for arbitrarily

* The term "canonical" form means "standard" or "unique" form.

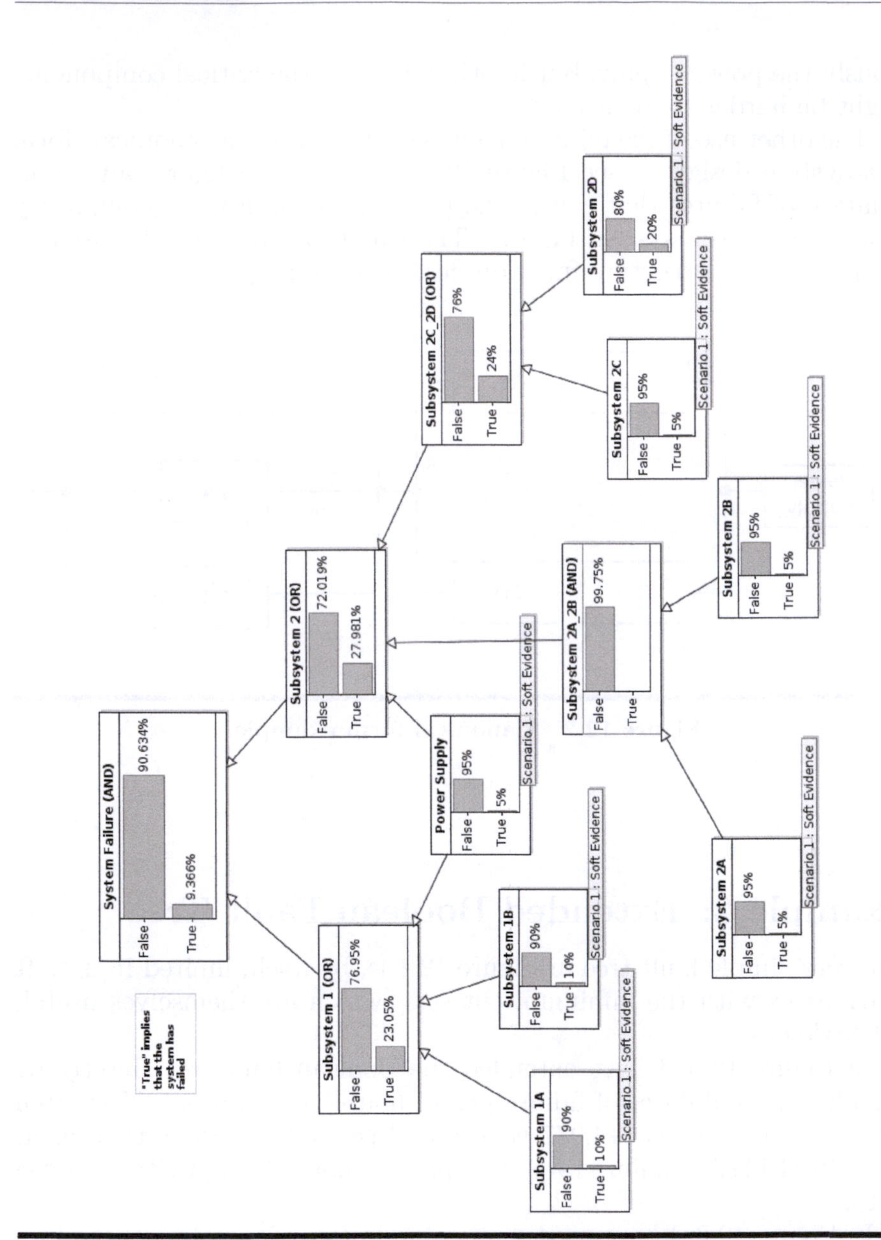

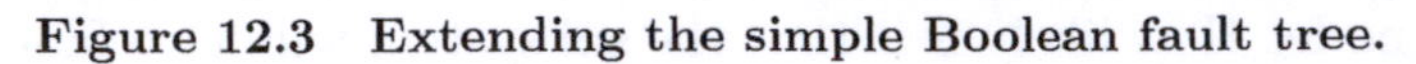

Figure 12.3 Extending the simple Boolean fault tree.

complex distributions, or even empirically measured distributions for which no closed-form solution exists. Figure 13.2 on page 186 shows a tree (not in that case a fault tree) with more complex distributions.

Example 3: Bayesian Fault Tree

Appendix B describes the general characteristics of a Bayesian network and that information is not repeated here. Rather, this section describes those aspects of a Bayesian network that are particularly useful for expressing fault trees. Reference [1] provides a more thorough account of using Bayesian networks for expressing failure conditions, in particular, using more complex failure distributions than I use here.

The first advantage of a Bayesian network is the introduction of noisy conjunctions (particularly noisy OR and, to a lesser extent, noisy AND). These are described below and illustrated in Figure 12.5. The second advantage is being able to exploit Bayes' theorem to reason "backwards" by adding an observation to an intermediate node in the tree as is illustrated in Figure 12.6.

Noisy OR

As described above, the noisy OR conjunction in the expression

$Z = X_1$ noisy OR X_2

states that:

- Generally speaking, if X_1 or X_2 fails, then Z fails.
- However, the individual failure of X_1 or X_2 does not *always* cause Z to fail — there is a probability associated with each.
- Even if neither X_1 nor X_2 fails, Z might still fail because of some cause we have not considered. This is known as the "leakage."

Formally, noisy OR is defined as follows:

$$P(Z|X_1, X_2, \ldots, X_N) = 1 - (1 - P(l)) \prod_{i=1}^{i=N} (1 - P(Z|X_i)) \qquad (12.1)$$

where $P(l)$ is the leakage probability, and $P(Z|X_i)$ is the probability of Z occurring if X_i occurs and none of the other $X_j, j \neq i$ occurs (that is, it is the level of influence that X_i by itself has over Z). This definition assumes that the X_i are causally independent — that is that if one of them occurs, that does not alter the probability of any of the others occurring. Reference [2] provides some justification for this definition of noisy OR.

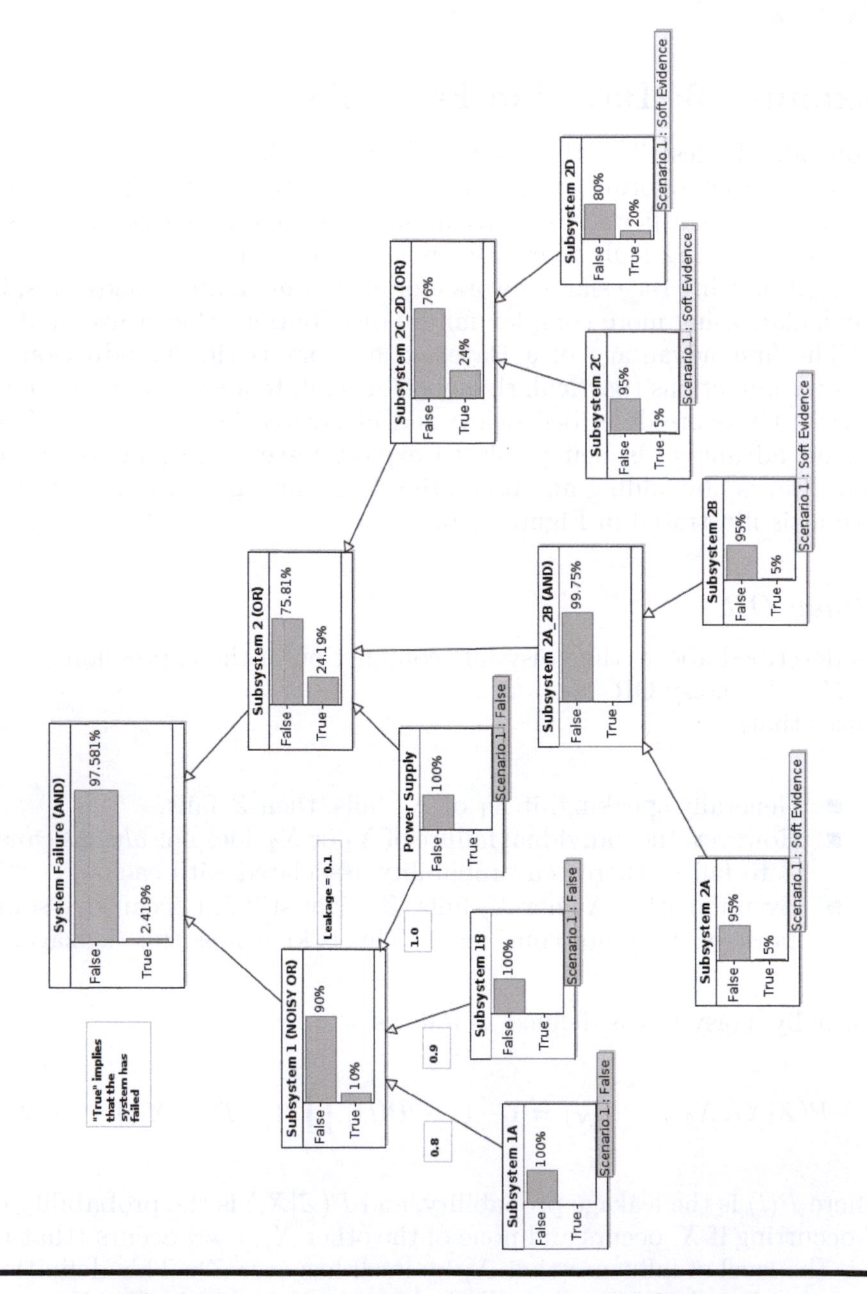

Figure 12.4 Noisy OR — illustrating the leakage.

The leakage parameter is illustrated in Figure 12.4, where the failure of Subsystem 1 is now represented as a noisy OR with a leakage of 0.1. As can be seen, even if the power supply and Subsystems 1A and 1B are 100% false (no failure), there is still a 10% chance that Subsystem 1 has failed.

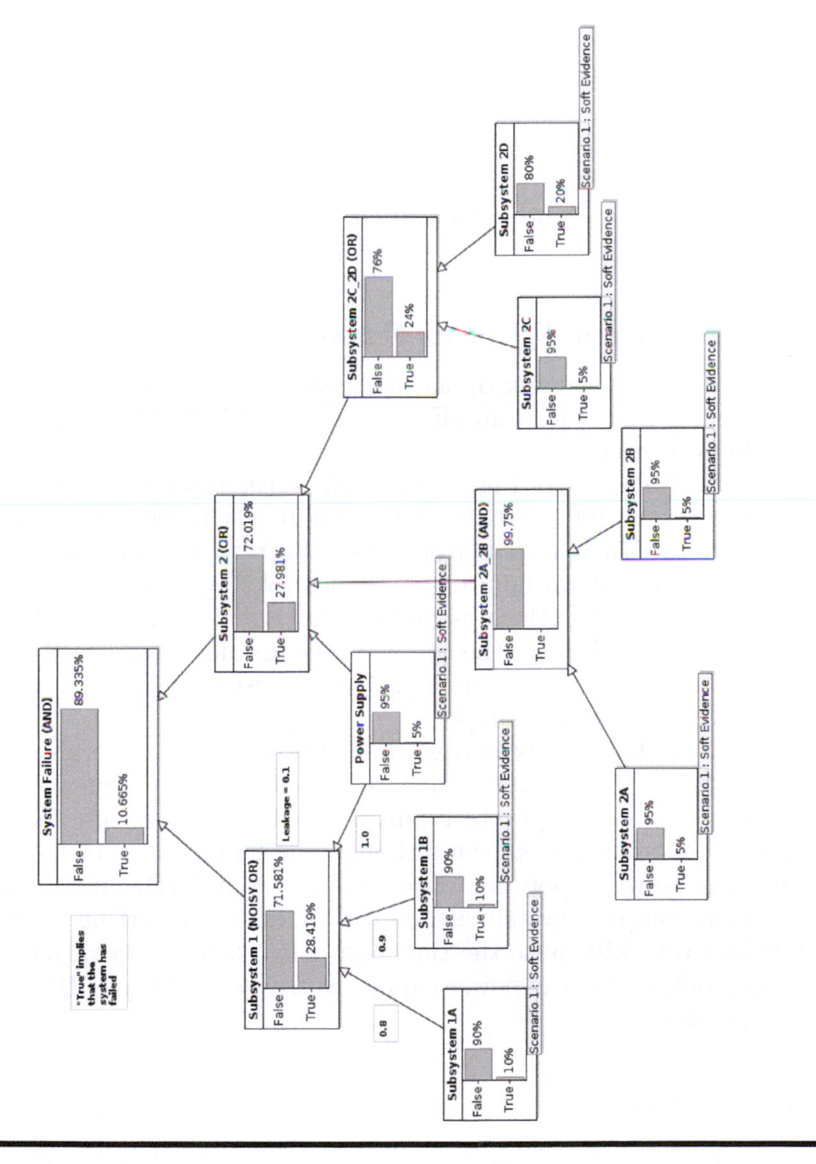

Figure 12.5 Noisy OR.

Figure 12.5 illustrates the integration of the leakage with "soft" values (i.e., values other that 1.0 (true) and 0.0 (false)) associated with the probability of failure of the components of Subsystem 1.

A weight of 1.0 has been assigned to the power supply: a Boolean condition saying that the failure of the power supply *always* causes the failure of Subsystem 1. If all the weights are 1.0 and the leakage is set to 0.0, a noisy OR reverts to a Boolean OR.

Weights of 0.8 and 0.9 have been associated with Subsystems 1A and 1B respectively. This indicates that if Subsystem 1A fails, even if Subsystem 1B and the power supply are operating, Subsystem 1 will fail 80% of the time.

Comparing Figures 12.5 and 12.3, it can be seen that introducing the noisy OR has slightly changed the overall failure characteristics of the system.

Reasoning from Effect to Cause

As described in Appendix B, one of the strengths of a Bayesian network is its ability to argue from effect to cause: "X has happened; what probably caused it?"

This is illustrated in Figure 12.6, where the top-level system failure has been set to 100% true — the system *has* failed. The Bayesian algorithm can then reason from that effect to the probable causes. Clearly, as the system failure is a Boolean AND, system failure being true indicates that both Subsystem 1 and Subsystem 2 must be 100% true. But, beneath that, the probabilities are not completely obvious. However, as expected, the most probable cause of the system failure is the failure of the power supply.

Reasoning backwards in this way is probably most useful when the failure rate of part of the system is known from experiments or field experience. For example, in Figure 12.5, it might be that the failure rate of Subsystem 1 is known although the failure rates of its components (the power supply and Subsystems 1A and 1B) and the failure rate of the complete system are, as yet, unknown. Inserting Subsystem 1's failure rate will cause the Bayesian network to be recalculated and values produced for the power supply, Subsystems 1A and 1B, and the overall system.

Noisy AND

In addition to noisy OR, there are other noisy conjunctions and operators; see reference [3] for a full description. Noisy AND is constructed in a manner analogous with noisy OR in that all inputs must be true

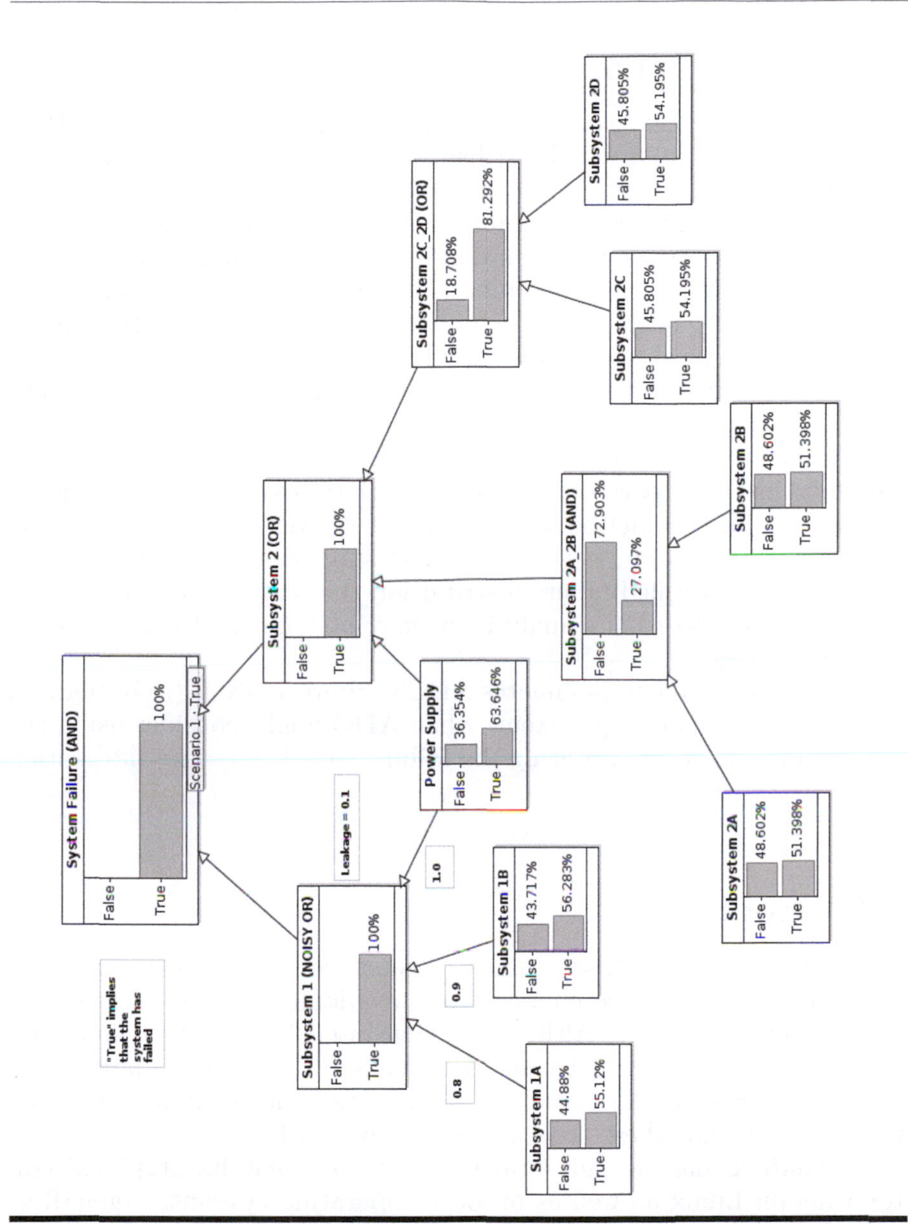

Figure 12.6 Reasoning from effect to cause.

to a particular degree for the output to be true, and even then there is a possibility (the leakage) that the output is not true.

Combining FTAs

Taking our imaginary companies from Chapter 4 as an example, Beta Component Incorporated (BCI) has to deliver a component to Alpha Device Corporation (ADC) for integration into ADC's device.

ADC will need to perform a failure analysis for the complete system and will therefore need information regarding the failure modes and frequencies of all bought-in components, including the one from BCI. For open-source components, such information may be hard to acquire, and ADC may need to do some significant work either assessing failure modes or ensuring that the system design takes frequent component failures into account.

For commercial components, such as that from BCI, ADC will presumably insist as part of the commercial terms of purchase, that a validated failure model be included in the delivery. Depending on the commercial relationships, the component supplier might be required to present only a verified failure distribution that can be incorporated as a single node into ADC's analysis, or may be required to deliver a full failure analysis.

If BCI's component dominates ADC's failure analysis (as it might in the case of an operating system), then ADC might consider using the same tool as BCI for capturing the failure model to reduce integration costs.

FTA Tools

I have used the AgenaRisk product to generate the figures in this chapter. This is one of the commercial tools available for Bayesian analysis, Netica being another. Although I have used both Netica and AgenaRisk successfully for FTA and safety-case preparation, I have never carried out a systematic analysis of the available tools and certainly cannot claim that these are the best on the market.

AgenaRisk has the advantage over Netica that its graphical editor runs on Linux as well as on other operating systems. AgenaRisk will also perform dynamic optimization of the discrete levels used to describe a variable. This is described in detail in Appendix D of reference [4], but, in summary, the problem it helps to solve is choosing the number of discrete intervals into which to divide a continuous distribution. If the analyst chooses too many small intervals, then the computation will be more precise, but could take a very long time to complete. Choosing too few, coarser intervals reduces the computation time at the cost of a less precise result. Selecting an optimal quantization level is difficult, and AgenaRisk's dynamic optimization takes this chore away from the analyst.

Netica has the advantage over AgenaRisk in that its graphical editor is completely separated from its calculation engine. This means that the graphical editor can be avoided and the model built programmatically using C, Java, C#, Visual Basic, or C++. Libraries are provided to allow the model to be passed directly to the calculation engine. This avoids the use of the Netica graphical interface, which has the disadvantage of running only on Microsoft's Windows Operating System. With AgenaRisk, in contrast, there is no published application programming interface (API) for the calculation engine; everything has to be entered using the graphics editor.

This is one area where I have never found a really good open-source alternative to the commercial tools.

The Use of FTA

FTA is a useful technique for expressing the failure model of a software-based system. The noisy conjunctions, particularly noisy OR, that are available with Bayesian representations allow subtle relationships to be expressed.

All FTA relies on an estimation of the software failure rate, and most companies have excellent sources of that information in the code repository, the problem report database, and the code review reports.

References

1. D. Marquez, M. Neil, and N. Fenton, "A new Bayesian Network approach to Reliability modelling," in *5th International Mathematical Methods in Reliability Conference (MMR 07)*, 1–4 July 2007.
2. F. G. Cozman, "Axiomatizing Noisy-OR," in *Proceedings of the 16th European Conference on Artificial Intelligence (ECAI-04)*, 2004.
3. Norsys, "Noisy-OR, -And, -Max and -Sum Nodes in Netica," *Norsys Net Library*, 2008.
4. N. Fenton and M. Neil, *Risk Assessment and Decision Analysis with Bayesian Networks*. CRC Press, 2013.

Chapter 13

Software Failure Rates

Measure what is measurable, and make measurable what is not.

Galileo Galilei

In order to carry out any form of failure analysis, whether Markovian, fault tree analysis (FTA), or failure mode, effects and criticality analysis (FMECA), it is necessary to have an estimate of the failure rates of the components.

Underlying Heresy

It must be said at the outset that the contents of this chapter are considered heretical by many people in the world of safety-critical systems. IEC 61508 makes the assumption that, whereas hardware failures can occur at random, software failures are all systematic, representing problems with the design, rather than problems with the implementation. This belief is founded on the principle that hardware wears out, but software doesn't.

This is an extremely naïve point of view that originated in a world of mechanical hardware and single-threaded programs running on single-core processors with a simple run-to-completion executive program rather than a re-entrant operating system. It also makes the assumption that the hardware of the processor will always correctly execute a correctly compiled program.

These assumptions are not true of the integrated hardware of today's software-based systems.

The C program listed in Figure 5.4 on page 70 is by itself sufficient to dispel the belief in non random software failure. That program *sometimes* fails with a floating point exception. The probability of failure depends on many factors, including whether the processor on which it is running is multicore or single-core and what other programs are being executed on the processor at the same time. It is impossible to predict whether it will fail for a particular execution, but given a stable environment, a statistical estimate can be made.

This is very similar to the failure of a mechanical hardware component — it is not possible to say whether a spring will fail the next time it is used, but given the history of similar springs, it is possible to provide a statistical estimate of its failure.

Anecdote 15 *I once submitted a paper to a journal containing a slightly modified version of the program in Figure 5.4. The referee to whom the paper was assigned remonstrated that I had made a mistake and the program as written could not fail. I convinced the editor that I was right, and the paper was published. On the very day that the journal appeared, I received an email from a safety engineer working for a large company, telling me that I had made a mistake. I convinced him that he was wrong and the editor suggested that he submit a "letter to the editor," which I could then refute. He declined to expose his mistake.*

If this extremely simple, two-threaded program can be studied ("code inspected") by reputable engineers and the fault not detected (see Anecdote 15), then what chance do we have with a more typical program running in a safety-critical embedded device with dozens or even hundreds of threads? On page 223, I describe the tool necessary to demonstrate that this program *can* fail, and also determine a sequence of events that can lead up to that failure.

There is a slightly different argument that can be applied to statistical failure of software. The reason why we cannot predict whether a mechanical spring will fail the next time it is used is not because it is, by its nature, unpredictable, but simply because we don't know enough. If we knew the position and velocity of every molecule in the spring, had a complete characterization of its environment (temperature, etc.) and had a large enough computer to do the calculation, we *could* in principle (to within quantum effects!) predict whether it will fail next time it is used. The apparent randomness of the hardware failure is caused by the enormous number of states in which the spring may be and our consequent inability to perform the calculation. And the same can be said for the program in Figure 5.4: If we knew the state of every thread in the the program's environment and had sufficient computational power, we could determine precisely whether it will fail during a particular execution.

Unfortunately, when combined with the libraries and operating system, it is easy to show that the number of possible states in which the program in Figure 5.4 can be exceeds 10^{100} — many more than the number of nucleons in the universe.

Whether both the software and hardware failures are genuinely random, or whether they are both predictable but appear random because of our ignorance, this argument puts today's multi-threaded software into the same class as mechanical hardware, and the type of techniques used in hardware failure analysis become available to software.

Compiler and Hardware Effects

There is one more nuance that can be extracted from Figure 5.4 and I am grateful to my colleague, Daniel Kasza, for pointing it out. Without the `volatile` keyword in the declaration of the global variable `x`, the compiler is free to remove the loop completely and "silently" replace it with something like `x = x + 100;`. For example, if the program is compiled for an ARM processor with the `-Os` command-line option, the assembler that is issued is as follows:

```
addOne():
    ldr r3, .L2
    ldr r2, [r3]
    adds    r2, r2, #100
    str r2, [r3]
    bx  lr
.L2:
    .word   .LANCHOR0
x:
```

The architecture of the processor can also affect the execution of the code generated by the compiler: Does the processor on which the program in Figure 5.4 is to be run guarantee that the whole of the integer `x` is loaded and stored atomically? It may be that the underlying hardware loads x into a register in two parts, potentially allowing other accesses (writes) to occur between the two parts of the load operation.

This will bring additional perceived randomness to the program. The compiler, particularly when invoked to provide time or space optimization, can change the intended behavior of a program, and the hardware on which the program is run does not necessarily guarantee correct execution of the instructions that the compiler has generated.

Assessing Failure Rates

Even if we accept that software failure in a tested and shipped product is normally caused by Heisenbugs and that, by definition, those failures are statistical, we still have the problem of assessing the failure rate of a particular software module.

Large hardware companies keep a detailed history of return rates, and smaller companies use some of the publicly available tables (e.g., the standard British Handbook of Reliability Data for Electronic Components used in Telecommunication Systems — HRD5). These tables list every type of component (resistor, capacitor, inductor, etc.) and provide rates for each type of failure (resistor going short-circuit, resistor going open-circuit, etc.). Such values can be used in a FMECA to estimate the overall dependability of the system.
I recount my "eureka" moment, based on hardware failure prediction, in Anecdote 16.

Anecdote 16 *I was working for a telecommunications company when a hardware dependability engineer suggested that we look at the return rates of a product that had been in field for a few years and compare the actual failure rate with that predicted by the failure analysis completed before the product was shipped.*

We found that the prediction had been extremely accurate: to several decimal places! This demonstrated the utility of performing hardware failure analysis and justified the time spent on it.

We then investigated more deeply and found that a whole batch of devices had been dropped in a warehouse and destroyed by a truck. This opened my eyes to the real meaning of failure rates — hardware failure rates are not calculated from the physics of carbon-film resistors, they are based on history. Over the years, the probability of a unit being dropped in a warehouse had become incorporated into the failure figures, and this is what made those figures so accurate.

And the one thing that software teams have is history.

With the possible exception of startups with no corporate history, all software companies have similar databases for their software. The code repository contains details of every change made to every module, the code review repository contains details of who reviewed the changes to the module and how many changes had to be made before the module was acceptable, and the problem database ties those changes to bug fixes or product enhancements.

This allows a level of analysis to be performed that hardware engineers can only dream about: "When Joe writes a module in C of 500-750 non comment lines in the XYZ subsystem, then a follow up change has to be made within 3 months in 73% of cases to fix a bug." The information available even allows the probability distribution to be determined.

This information is not used to castigate Joe for his poor design and coding — it may be that Joe is the senior person in the team and is always given the hardest parts of the design to implement. Rather, this information allows us to estimate quite accurately the probable failure rate of another piece of code in the XYZ subsystem that Joe is producing.

Fallacy 6 *It is not true that hardware failure analysis is always accurate. For example, reference [1] describes an experiment carried out by the European Power Supply Manufacturers Association (EPSMA). Sixteen member companies of EPSMA were asked to estimate the mean time to failure of a simple ten component, 1 Watt, DC-DC power converter using their normal failure modeling.*

The results were incredible, ranging from 95 to 11895 years (a ratio of 1:125). After some discussion between the companies, the results were reduced to the range of 1205 to 11895 years (still a ratio of 1:10).

This is not to laugh at hardware failure rate prediction — we know that it is useful. The moral of the story is that we can do at least as well for software.

This type of data mining of the source code repository and problem report database is very similar to that described in the discussion on fault density assessment on page 270 and the improvement of static code analysis described on page 266. The same data are being mined, using similar tools, but for different purposes.

Modeling the Failures

Given the basic failure information, there are many models available to estimate the failure rate of a product after release. These models typically involve some form of curve fitting to a predefined statistical distribution to determine the (unknown) number of bugs remaining in the software. Reference [2] by Norman Fenton *et al.* points out a major flaw in this technique. A particular software vendor installed company-wide metrics to determine the best code, the code from which other project teams could learn. Once all the metrics had been gathered and the best projects identified by having extremely low bug reporting rates, someone pointed out that all of the "best" products had been flops. As no one was using them, the bug reporting rate from the field was zero.

Having identified this logical flaw (a piece of software that is never used results in no bug reports), reference [2] goes on to provide a representation, based on Bayesian Networks (see Appendix B), where the amount of usage is taken into account. A somewhat simplified version of the model given reference [2] is included as Figure 13.1, where I have deliberately selected an extreme example: a very complex project with very poor design processes, a very low level of verification, and a lot of field usage. A more realistic example is given in Figure 13.2, where it can be seen that the estimated number of shipped bugs is much lower, the dotted line in each box representing the mean.

It can be seen from Figure 13.1 that slightly more sophisticated distributions are used than were applied to the fault tree analysis (FTA) in Chapter 12. In summary, it is based on the assumption that the number of residual defects is dependent on the number of defects put into the code during development and the number of defects found and removed during pre-shipment testing. The number and frequency of defects appearing in the field depend on the number of residual defects and the amount of operational use, thus avoiding the problem identified above of low use generating no bug reports.

Initially, the prior probabilities of the Bayesian network (see Appendix B for the meaning of the term "prior") come from mining the company's history, as described above. As the project progresses, the

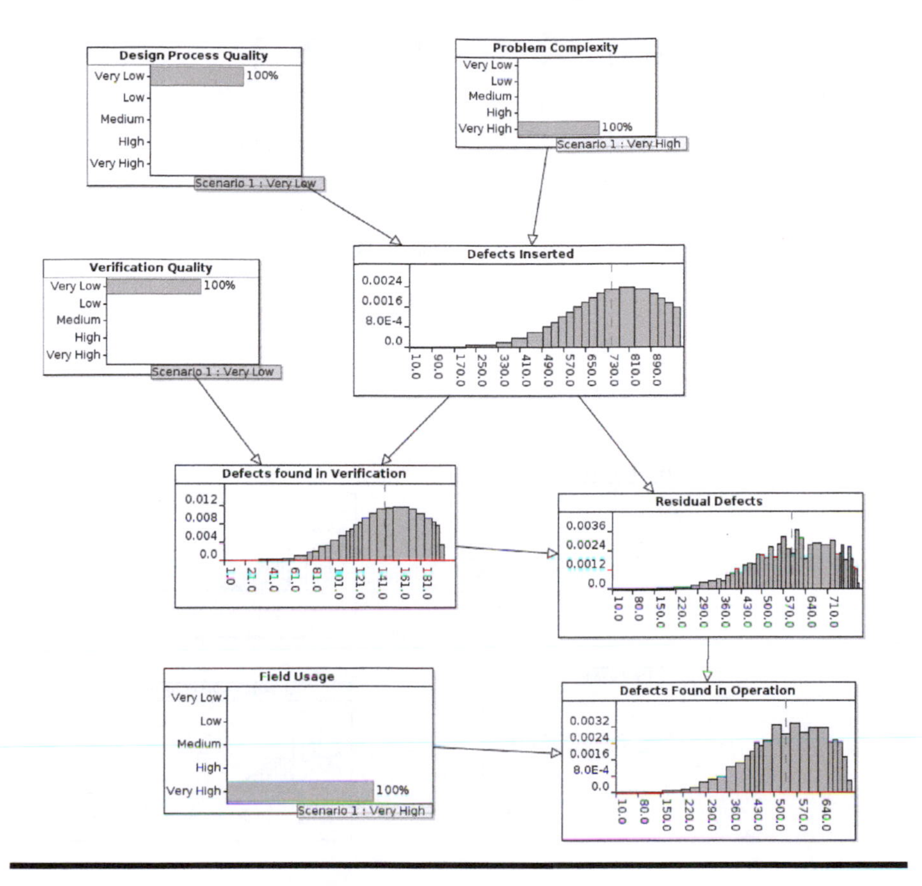

Figure 13.1 Estimating the software defects: Extreme case.

measured number of defects found during testing and the level of operational use can be added to the computation.

It is reported that this technique has been used by companies, such as Siemens (reference [3]) and Philips (reference [4]).

References

1. European Power Supply Manufacturers Association, "Guidelines to understanding reliability prediction," 2005.
2. N. Fenton, M. Neil, and D. Marquez, "Using Bayesian networks to predict software defects and reliability," 2008.
3. H. Wang, F. Peng, C. Zhang, and A. Pietschker, "Software Project Level Estimation Model Framework based on Bayesian Belief Networks," in *QSIC '06: Proceedings of the Sixth International Conference on Quality Software*, (Washington, DC, USA), pp. 209–218, IEEE Computer Society, 2006.

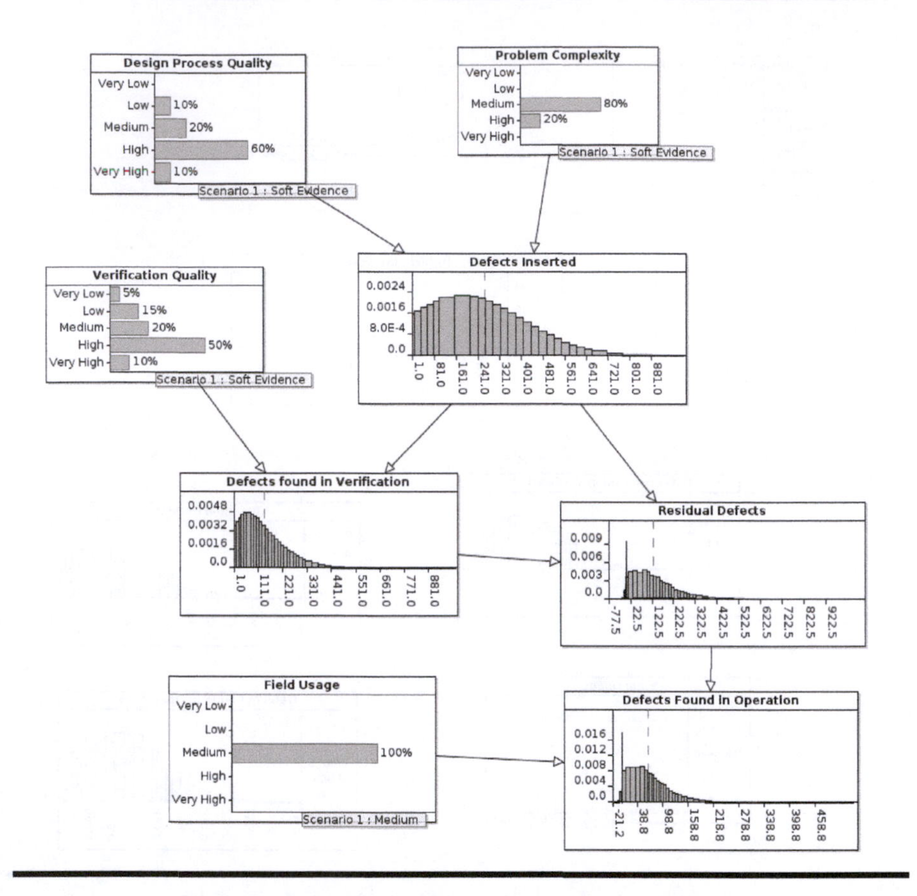

Figure 13.2 Estimating the software defects.

4. N. E. Fenton, P. Krause, and M. Neil, "Software Measurement: Uncertainty and Causal Modelling," *IEEE Software*, vol. 19, no. 4, pp. 116–122, 2002.

Chapter 14

Semi-Formal Design Verification

Together with Chapter 15, which covers formal verification methods,* this chapter considers some of the techniques that can be used to answer the question: "Will a system implemented to this design deliver what is required of it?" More detailed questions might be, "What confidence can I have that this system will meet its dependability requirements? How confident can I be that the protocol between the two subsystems can never lock up? How confident can I be that the system will meet its real-time requirements under *all* combinations of events?"

After a short explanation of when it might be necessary to carry out design verification retrospectively on a system that has already been implemented, this chapter describes the semi-formal techniques of discrete event simulation (see page 190) and timed Petri nets (page 199). These techniques are not independent and can be seen as different languages for expressing the same simulation.

Nor are these the only semi-formal techniques used to verify designs; Table B.7 of IEC 61508-3 includes references to data flow diagrams, UML diagrams, decision tables, message sequence charts, entity-relationship-attribute models, finite state machines, and other semi-formal methods. I have selected discrete event simulation and timed Petri nets for examination here because I have found them to be less well-known than these other techniques. They are also supersets of some of the other notations, in particular of finite state machines, mes-

---

* For a description of the difference between informal, semi-formal, and formal methods, see page 17.

sage sequence charts, SysML's activity diagram, and data flow diagrams.

Anecdote 17 *I once worked on the verification of a design for a very sophisticated thread scheduler. I tried for some time to produce a formal verification of its correctness, but I could not complete the proof — the integer arithmetic within the algorithm was too complex for the automated theorem provers. However, by applying a discrete event simulation to those parts of the algorithm that were not amenable to formal proving, I was able to demonstrate the correctness of the algorithm with high confidence.*

The use of back-to-back comparison between a model and code described on page 284 is also an example of combining one of the techniques described here, discrete event simulation, with a validation technique.

Verification of a Reconstructed Design

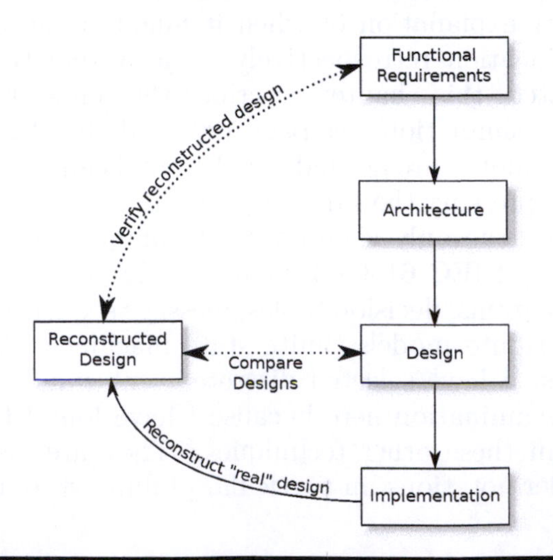

Figure 14.1 Retrospective design verification.

Ideally, a design is verified as it is produced; once some of the design decisions have been made, those decisions can be verified to confirm that they meet the system's requirements.

However, there are times when design verification needs to be performed later, once the design has been implemented; see Figure 14.1.

The most obvious case for creating a design retrospectively is where an existing product, developed to a lower standard, now needs to be validated for use in a safety-critical system, and where the original design was inadequately recorded or has been lost.

There are also less obvious reasons for reconstructing the "design" from the implementation and verifying it. During product development, the links between the requirements, the architecture, and the design can be kept fairly tight. Each safety requirement can be traced through the architecture and design, and each facet of the design can be traced back to a requirement. See section 7.4.2 of ISO 26262-6:

> *During the development of the software architectural design the following shall be considered: ... the verifiability of the software architectural design; ...*
> *NOTE: This implies bi-directional traceability between the software architectural design and the software safety requirements.*

However, this close tracing can be lost in the step between design and implementation — the implementation team is likely to be large and the implementation process can be harder to control. Rebuilding from the implementation the "design" that has actually been built and comparing that with the real design, as illustrated in Figure 14.1, can then be useful in closing the loop.

The Food and Drug Administration (FDA) in the USA has the responsibility for ensuring that new medical devices are safe for use in that country. During the first decade of the 21st century, the FDA realized that testing software-based devices was becoming increasingly ineffective, as Heisenbugs began to predominate* and cast around for other techniques. Reference [1] by Dharmalingam Ganesan *et al.* describes how design-level characteristics of an infusion pump system can retrospectively be derived from the source code. These characteristics can then be used to verify that safety requirements are met and can also provide useful pointers for deeper analysis of the implementation through static analysis.

---

* See page 21 for a description of Heisenbugs and their effect on test efficiency.

Discrete Event Simulation

When we are validating a design, discrete event simulation can answer some important questions:

- What level of dependability can I expect from a system implemented to this design? When used like this, simulation replaces or, more usefully, augments fault tree analysis — see page 166 — and Petri nets.
- How many resources of each type (memory, mutexes, message buffers, etc.) will be needed to ensure that the probability of exhaustion is below x per hour of operation?
- What delays can be expected, given particular loads on the system?
- How will complex algorithm X behave under extreme conditions? When used in this mode, discrete event simulation can complement formal verification — see Chapter 15.

Digital computers have been performing discrete event simulations almost since their inception. On my bookshelf I have a manual for the general purpose system simulation (GPSS) language (reference [2] by Thomas Schriber) dated 1974, at which time GPSS was already a mature language. Appendix F to reference [2] now seems pleasantly quaint. It lists the maximum size of a model that can be executed, this being dependent on the amount of memory fitted in the mainframe computer running the GPSS interpreter. The amount of memory varies from 64 kbytes to a massive 256 kbytes.

Anecdote 18 *When I first met GPSS (and bought reference [2]), I was simulating the flow of domestic hot-water boiler panels through an automated paint shop. The simulation was running on an IBM 370 mainframe with a total of 128 kbytes of memory.*

Discrete Event Simulation and the Standards

Table 6 in ISO 26262-6 recommends simulation as a method for the "verification of the software architectural design." Simulation is also described as a modeling technique in Annex B of ISO 26262-6.

IEC 61508 still uses the term "Monte-Carlo simulation" for discrete event simulation. This was one of the very early terms, rarely heard in computing now.

Section B.6.6.6 of IEC 61508-7 calls up discrete event simulation when modeling requirements exceed the exponential distributions that can be handled by Markov models — see Chapter 11 — and the whole of B.6.6.8 is dedicated to Mont-Carlo simulation.

According to IEC 61508-7, discrete event simulation is cited as a recommended technique for failure analysis in Table B.4 of IEC 61508-3. However, this reference is incorrect — there *was* such a reference in the first edition of IEC 61508, but it was deleted in the second edition with the comment, "This technique has been deleted on the grounds that it has not proved useful in the development of safety related systems." I believe this to be a serious mistake and it is interesting that the technique still appears in Annex B of IEC 61508-6, where it is used in an example for evaluating probabilities of hardware failure.

For medical systems, ISO 14971 describes the use of discrete event simulation for estimating probabilities (section D.3.2.2) and advises that it be used together with other approaches, each working as an independent check on the others.

Pseudo-Random Number Generation

All forms of discrete event simulation need a stream of random numbers. Numbers that are truly random are both hard to generate (requiring, for example, a counter of atomic disintegrations) and difficult to use because they make it impossible to run the same simulation twice for debugging or comparison purposes.

Rather than genuinely random numbers, pseudo-random numbers are therefore generally used. A well-known quotation from reference [3] by John von Neumann is that "Anyone who considers arithmetical methods of producing random digits is, of course, in a state of sin. For, as has been pointed out several times, there is no such thing as a random number — there are only methods to produce random numbers, and a strict arithmetic procedure of course is not such a method."* Pseudo-random numbers are generated deterministically from an original "seed." Given the same seed twice, the same sequence of pseudo-random numbers will be generated.

If a source of such numbers is to be used in a simulation to provide

---

* This reference is very hard to find. At the time of writing this book, a scanned copy can be found at `https://dornsifecms.usc.edu/richard-arratia-usc/hard-to-find-papers-i-admire/`.

evidence for the correct operation of a safety-critical system, some attention will have to be given to demonstrating that the pseudo-random number stream is adequately random. Over the years, many generation algorithms have been tried which have resulted in either very short runs of numbers before repetition occurs or poorly distributed outputs. Chapter 3 (volume 2) of reference [4] by Donald E Knuth is venerable, but still a reliable source of information about checking the randomness of a stream of generated numbers.

Simulation of Deterministic Systems

Simulations can be used to analyze both deterministic and nondeterministic systems. A good example of the former is the exercise of estimating π by simulating the throwing of darts at a dart board; see Figure 14.2. If the position of each dart is determined by two random numbers, each in (0, 1), then the fraction of the darts that fall onto the circular dart board is $\frac{\pi}{4}$.

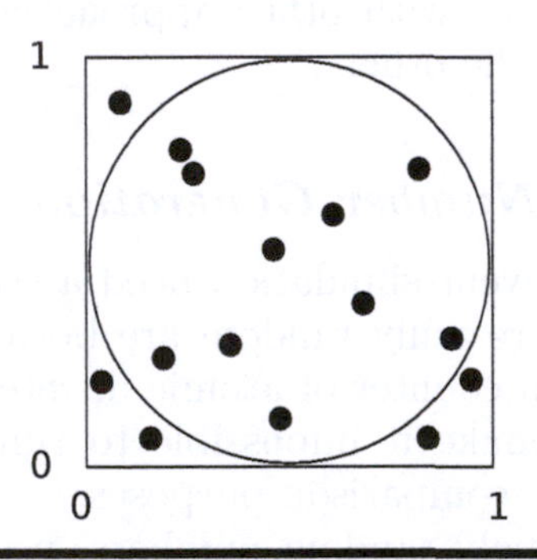

Figure 14.2 A dartboard and random dart thrower.

I have just executed a simple python program to throw 10,000,000 darts at the board. 7,853,379 fell onto the circular dart board, and 2,146,621 fell outside it. This gives an estimate of $\pi \approx 3.14135$. Of course, this exercise is really measuring the randomness of python's pseudo-random number generator rather than being a practical method for estimating π.

In this case there is nothing random about the system being simulated, pseudo-random numbers are used only to explore it. While calculating π in this rather inefficient fashion isn't very useful, the technique of simulating a deterministic system is often useful when a mathematical result (e.g., a high-dimensional integral) cannot be obtained analytically or when it is impossible to set up the equations describing a complex system. In the latter case, the pattern illustrated in Figure 14.3 can be used.

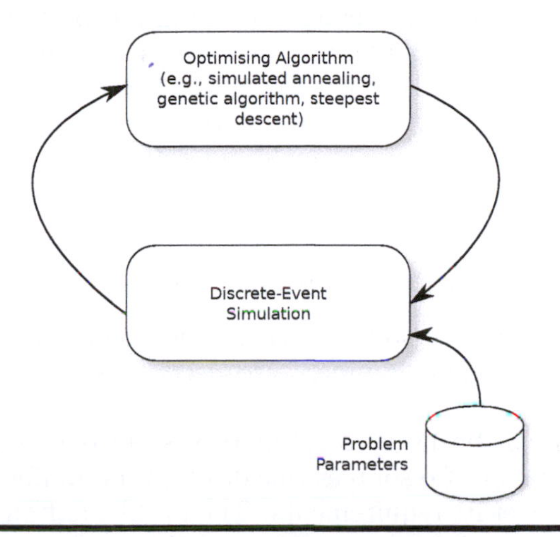

Figure 14.3 Simulation as an optimization aid.

Here, some optimizing algorithm is trying to find the minimum (or maximum) value of a complex, multidimensional surface representing the physical system. Neither IEC 61508 nor ISO 26262 mentions optimization algorithms, and so I will not delve into them here. Suffice it to say that simulated annealing, genetic algorithms, and other minimisation techniques work by probing the multidimensional surface of the parameter being minimized (e.g., cost) and working their way "downhill" toward a local minimum. They need some mechanism for answering the question, "If we were to do x, what would be the cost?" In some cases, this can be answered by a simple arithmetic calculation; in other cases a discrete event simulation may be the simplest route.

Consider the automated paint shop in Anecdote 18 above. The problem was not that the progress of boiler panels through the acid bath, electro-static paint room, and the oven was random. On the contrary, this was very deterministic. The problem was that it took a panel something like 40 minutes to go through the process, and it took over an hour to change paint color (blowing out the old paint and filling the system with the new color). On a Friday afternoon, the number of panels of different colors for the following week was known, and what was required was the optimal painting schedule. A boiler couldn't be built until all its panels (say, 3 red, 4 blue, and 1 white) were available and, unfortunately, the storage area for painted, but not yet used, panels was limited. Given a large number of different boilers to be built and the long delay in changing paint color, finding a schedule to maximize the efficiency of the paint shop was not trivial.

I devised an optimizing algorithm (using steepest descent), but it needed to know what would happen were the schedule to be changed so that the white paint was kept flowing until 11:00 on the Tuesday rather than 09:00. This was excellent work for a deterministic simulation: simulating the panels going through and returning the week's results to the optimization algorithm.

Simulation of Nondeterministic Systems

In a nondeterministic system, such as one where requests arrive at random, pseudo-random numbers are used to simulate parts of the system itself.

Simulations can be used to estimate system dependability and to estimate the number of resources (file descriptors, buffers, etc.) needed by a system to meet its requirements. The problem often arises because of the word "estimate" in that sentence. Other techniques for assessing system dependability, such as Markov models or fault tree analysis, provide a precise (although not necessarily accurate) answer; simulation provides only a statistical assurance. A simulation may, for example, provide us with a 95% confidence that the system, as simulated, will not require more than 48 message buffers more often than once per week of continuous use.

This means that the simulation analyst needs to have a basic knowledge of statistical techniques, but this is an increasing need throughout design verification. In the past, when processors and software were simple, analysts looked for absolute guarantees on software termination: Under any sequence of events, process A will always complete within x milliseconds. With modern processors incorporating data and instruction cacheing, this is no longer possible, and "guarantees" become statistical: Under the anticipated sequence of events, the probability of process A not completing its actions within x milliseconds is less than y %.

Discrete Event Simulation Frameworks

For the simplest simulations, particularly those written to estimate system availability, no framework is really required; a program can be written directly to carry out the simulation.

For more complex simulations, one of the many commercial and open-source frameworks can be used. These divide into two main classes:

1. **Event-based simulation.**
   Eponymously, the focus here is on events. One event occurring (e.g., the failure of a component) may schedule one or more events for the future (e.g., the arrival of a repair technician). The simulation framework keeps a sorted list of future events and loops, pulling the next event from the list, updating the simulation time to the appropriate time (hence the term *discrete* event simulation — time moves in discrete steps), and processing it, adding new events to the list as required. The simulation finishes either when the event list is empty or the predefined simulation time is reached.
2. **Transaction- or process-based simulation.**
   This technique focuses on transactions that are born, move through their "life," interacting with each other and various system services, and are then terminated. A transaction might be a message received by a device that is queued while waiting for the processor and which then absorbs some of the processor's time before moving through other parts of the simulated system and finally being destroyed.

   In this case, two lists need to be maintained: the future events list, as with event-based simulations, and also a current events list, which holds transactions that are waiting for something to happen (e.g., a processing core to become ready). The efficiency with which the current events list is handled is often what determines the run time of a simulation.

I have found event-based simulations to be very easy to program for small systems, but have found that the number of events grows wildly as the complexity of the system increases.

Transaction-based frameworks, such as SimPy* and the venerable GPSS, provide abstractions for the types of entity with which transactions interact: queues, facilities that service queues (e.g., a processor), levels for which a transaction may wait (e.g., an amount of free memory), gates which may temporarily block the progress of a transaction until another transaction opens the gate, and many others.

---

* `http://simpy.readthedocs.org/en/latest/`

Discrete Event Simulation Example

Rather than further discuss discrete event simulation in the abstract, I shall demonstrate applying it to an example problem. Simulation can be used to estimate many things, but possibly the simplest thing to simulate is a system's failure rate. This generally does not involve any of the more sophisticated features listed above: queues, facilities, etc.

Figure 12.1 on page 168 illustrates a tiny part of a system, and the failure rate of this system has already been calculated by means of a Markov computation starting on page 158. One of the drawbacks of the Markov model is that it can only handle negatively exponentially distributed failure interarrival times. The advantage of a simulation is that any type of distribution can be used, even an empirical (measured) one where the mathematical distribution is unknown.

I give below the results of two simulations: the first using the Markovian assumption, the second with more complex distributions. The advantage of using the Markovian assumptions is, of course, that we already know the answer — see Table 11.1 on page 161.

Repeating the Markov Calculation

I wrote a short Python program to simulate the system moving through the states shown in Figure 11.2 on page 159 and, for the first example, assumed a negatively exponential distribution of fault and repair interarrival times — i.e., assumed the failures to be in accordance with the Markov assumptions.

Table 14.1 lists the results of the mean time spent in states 1 (no failures) and 12 (system failed) during each of 10 simulation runs, using different pseudo-random number seeds. Each run simulated 500 years of the system's life and took about 9ms to execute on my desktop computer. The numbers in Figure 14.1 can be compared with the correct value of 92.963044% and 0.063491% for states 1 and 12, respectively.

Although we have a precise answer from the Markov analysis, we have only a statistical answer from the simulation, and it is even clearer in Table 14.1 than it was in Table 11.1 that giving a large number of decimal places is completely meaningless.

Estimating the Confidence Intervals

Given the sample mean and variance, it is possible using normal statistical techniques to calculate a confidence interval:

$$\bar{x} \pm \frac{s}{\sqrt{n}} t_q(n-1) \tag{14.1}$$

Table 14.1 Simulation with Markovian assumptions.

Run	**Percentage of Time In ...**	
	State 1	**State 12**
1	92.5692035428	0.0547521172869
2	93.0931672238	0.0531698828013
3	93.5400195792	0.0600734354356
4	92.7074828523	0.0792807103518
5	92.8191917161	0.0537101528804
6	92.6779171393	0.061319266597
7	92.7161369847	0.0604424692454
8	92.7253859097	0.0832654911244
9	92.9228251701	0.0654410202069
10	92.9694212612	0.0693077010148
Sample Mean ($\bar{x}$)	92.8740751379	0.0640762246944
Sample Variance (s^2)	0.0791326184629	0.000108756153029

where $\bar{x}$ is the sample mean, s is the sample standard deviation, n is the number of samples, and $t_q(n-1)$ is the value of Student's t distribution corresponding to confidence level $q\%$ with $(n-1)$ degrees of freedom.

Given this formula, and reducing the number of decimal places to something sensible, the 95th percentile confidence interval on the percentage of time spent in State 12 (system failed) is

$$\bar{x} \pm \frac{s}{\sqrt{n}} t_q(n-1) = 0.064 \pm \sqrt{\frac{0.000109}{10}} \times 2.262 = 0.064 \pm 0.00747 \quad (14.2)$$

This gives a 95% confidence interval of 0.057 to 0.072, and the equivalent interval for State 1 (no failures) is 92.7 to 93.1. These ranges encompass the correct answers calculated from the Markov model.

When using this technique, remember precisely what the confidence interval means. It is not a confidence interval around the (unknown) correct value of the mean — it is a confidence interval around the next value that will be reported by the simulation. These may be very different things.

Confidence intervals may be calculated either, as here, by executing the simulation a number of times with different random number streams or by executing one very long run and breaking it into shorter pieces, treating each as independent of the others.

Simulating without the Markov Assumption

Of course, it is unnecessary to perform a simulation on this system when all the failure probabilities are Poisson, in that case the Markov model is completely accurate. The simulation becomes useful when the failure rate distributions are non-Poisson and, in particular, if they are measured distributions without a mathematical formula.

To illustrate this, I will keep the mean rates of the component failures and repair times the same, but will assume different distributions, as listed in Table 14.2. Note that all of the normal (Gaussian) distributions are truncated at 0.

Table 14.2 Assumptions on distributions.

Component	Mean Rate	Arrival Distribution
1A	0·333	normal, mean 3 years, sd 2 years
1B	0·333	normal, mean 3 years, sd 2 years
2A	0·333	uniform 2 to 4 years
2B	0·333	uniform 2 to 4 years
2C	0·333	uniform 1 to 5 years
2D	0·333	uniform 1 to 5 years
Power supply	0·2	normal, mean 5 years, sd 2 years
Repair, not failed	26·00	uniform 5 to 9 days
Repair, failed	365·25	uniform 14 to 34 hours

"sd" means "standard deviation."

Making the minor modifications to the simulation program to use the distributions from Table 14.2 instead of Poisson results in a 95% confidence interval for the proportion of time spent in State 1 (no failures) as 98.8% to 98.9%, and the time spent in State 12 (system failure) as 0.0088% to 0.0121%.

Note that, although the mean rates of failure and repair have remained the same for the two simulations, the mean up-time of the system has changed from 99.93% to 99.98%. Whether this is significant or not depends on the dependability requirements of the system.

Knowing When to Start

The example simulations given above are very simple and are not subject to the normal startup errors of more sophisticated simulations.

Normally, a simulation includes queues of transactions waiting for a server (e.g., queues of incoming packets waiting to be processed by the communications stack) and a number of servers. Initially, the queues are empty and the servers idle. Until the simulation has been running for some time, it provides an inaccurate model of the steady-state behavior of the real system.

Therefore, it is normal to allow the simulation to reach a steady-state and then reset all the statistics (queue lengths, server utilizations, etc.) and start collecting statistics again.

It is difficult to know when such a reset should take place: do it too early and the results will be incorrect, do it too late and processing time has been wasted.

Knowing When to Stop

If knowing when to start collecting statistics is difficult, knowing when to stop is even more difficult. In the example above, I simulated for 500 years. Why not 10 years? Why not 1000 years?

This is a particular problem if the system has rare events. If most events occur in the model every millisecond, but one particular event (e.g., a failure) occurs only every 5 years, then, to get statistical significance on the rare event, it will be necessary to run the simulation for a simulated time of perhaps 50 years. In that time there will be about 10^{12} occurrences of the millisecond event. It is the job of the modeler to avoid this situation.

The main practical criterion for deciding the length of the simulation run is monitoring key statistics and observing that they are no longer changing significantly.

Timed Petri Nets

Petri Nets and the Standards

IEC 61508-3 recommends the use of *semi-formal methods* in Table A.1 and refers from there to Table B.7 which contains a list of several methods considered semi-formal in the specification. These include timed Petri nets.

ISO 26262 does not mention Petri nets explicitly, but the reference to "simulation of dynamic parts of the design" in Table 6 of ISO 26262-

6 and its expansion in Annex B covers the use of Petri nets, which are suitable for this type of simulation.

The History of Petri Nets

Unlike many of the techniques described in this book, which seem to have grown out of some key ideas lost in the mists of history, Petri nets have a clear beginning. It is reported that in August 1939 the 13-year old Carl Adam Petri invented the notation to allow him to describe chemical processes. The idea was published in Petri's doctoral thesis (reference [5]) in 1962.

Since Petri's original design, a variety of Petri nets have appeared, including timed Petri nets, colored Petri nets, dualistic Petri nets, fuzzy Petri nets, prioritized Petri nets, and stochastic Petri nets.

For the purposes of embedded system design, timed Petri nets and their extension to timed stochastic Petri nets are the most useful kinds, and I will use the simple term "Petri net" to mean one of those.

What is a Petri Net?

A Petri net can be thought of as simply a different notation for describing a Markov model or a discrete event simulation. It is a graphical notation and is particularly useful for expressing the types of Markov models given in Chapter 11.

Reference [6] by Ajmone Marsan, although somewhat old, still provides an excellent elementary introduction to timed Petri nets.

A Simple Petri Net

The Petri net notation consists of places, transitions, tokens and an initial layout of tokens as illustrated in Figure 14.4. These are all abstract concepts and can represent different aspects of a system.

In that figure, places are represented by open circles, transitions by open rectangles (different types of rectangle represent different types of transition), and tokens by black dots. In this case, the initial layout has one token in place 1 and one token in place 4.

A transition can "fire" if each of its input places contain at least one token. When it fires, one token is removed from each of its input places, and one token is put into each of its output places.

I like to think of this as a type of board game to be played with the children during the holiday season when it's raining outside. Petri net tools allow the designer to simulate the transition of tokens, and many even call the option "the token game."

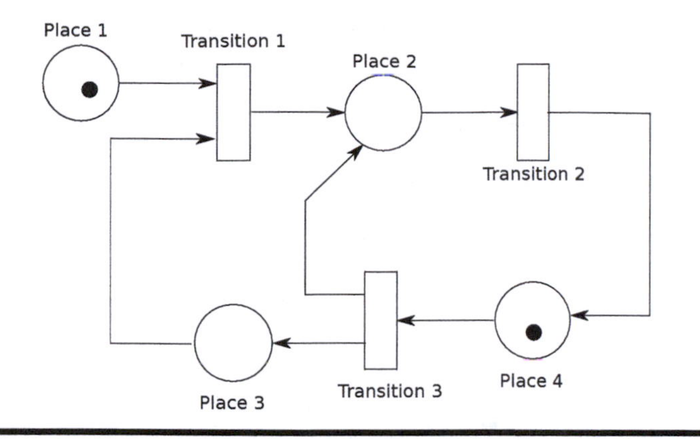

Figure 14.4 A simple Petri net.

In Figure 14.4, transition 1 cannot fire because, although there is a token in place 1, there is no token in its other input, place 3. Transition 2 cannot fire either, but transition 3 can. When transition 3 fires, the result is as given in Figure 14.5.

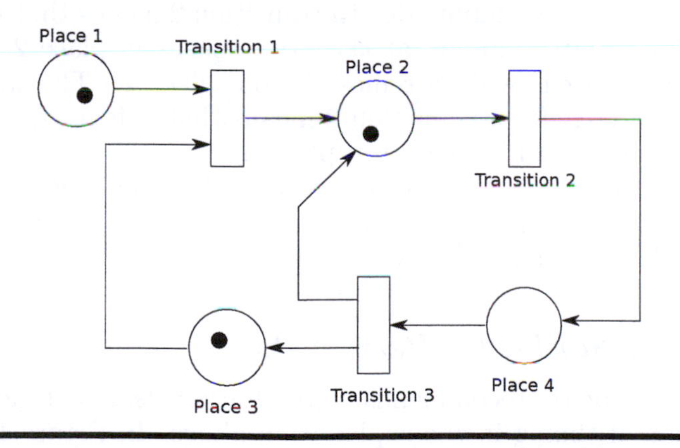

Figure 14.5 Figure 14.4 after a single step.

Note that there is no conservation of tokens: One token in place 4 has become two tokens, one in place 2 and one in place 3. Now transition 2 and transition 3 are able to fire, and the "game" progresses.

Extending the Simple Petri Net

There are a multitude of ways in which Petri nets have been extended. Figure 14.6 illustrates a few of the more common.

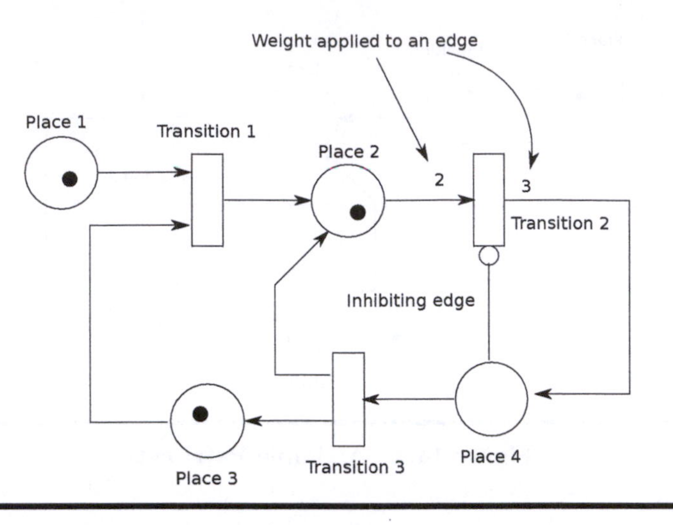

Figure 14.6 Simple Petri net extensions.

In that figure, weights have been introduced on two of the edges. The weight of 2 on the input edge to transition 2 means that transition 2 will not fire until there are at least two tokens in place 2. When it does fire, two tokens will be removed from place 2. The weight of 3 on the output edge from transition 2 means that, when that transition fires, three tokens will be placed in place 4.

Figure 14.6 also contains an inhibiting edge. This acts in the opposite way to a normal edge: A token in place 4 actually prevents transition 2 from firing, whatever the state of place 2.

Timed and Stochastic Petri Nets

IEC 61508 explicitly recommends *timed Petri nets*, and I glossed over the question of timing in the explanation above. In Figure 14.5, both transition 1 and transition 2 are able to fire. This raises two questions: Which will fire, and how long does a firing take? The first of these is answered below in the section on non-determinism (page 206). The answer to the second question is that for a timed Petri net, times (possibly zero) are associated with each transition. If the firing time is random in some way, then the network becomes a stochastic, timed Petri net.

Different authors use different symbols for transitions with different types of timing. Figure 14.7 lists the symbols that I normally use.

In a timed Petri net, a transition can be thought of firing in three stages:

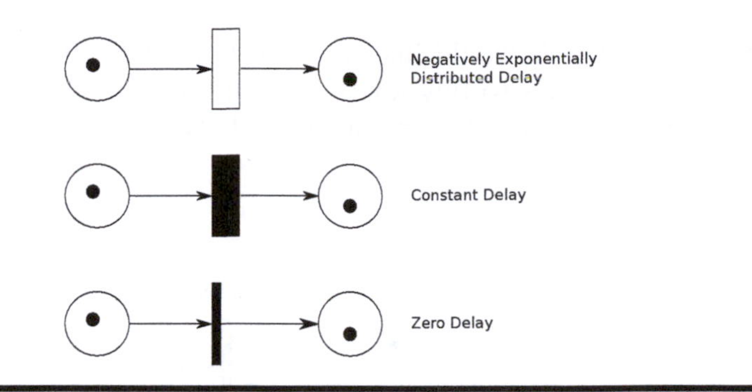

Figure 14.7 Transition types.

1. When enabled, the tokens are removed from the input places.
2. The delay associated with the transition takes place.
3. The tokens appear in the output places.

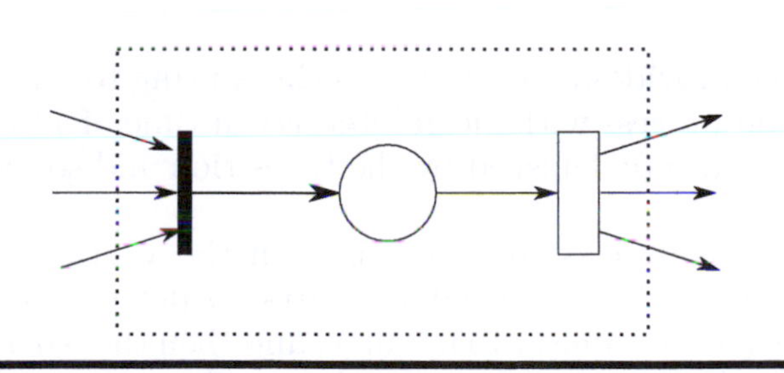

Figure 14.8 The internal structure of a timed transition.

This is illustrated in Figure 14.8, where an immediate (no delay) transition is used to remove the tokens from the input places and an intermediate place is used to hold the tokens internally until the delay has passed and they can appear in the output places.

The Barber's Shop

It is traditional in simulation circles to start with a barber's shop — this is the equivalent of the "hello world" program when starting to learn a new programming language.

Assume that people arrive to have their haircut in accordance with a Poisson process with a mean interarrival time of 14 minutes, wait in a waiting room until the barber is free, and are then served in a negative exponentially distributed time with a mean of 10 minutes. They then depart the shop. We would like to know how busy the barber will be and how many chairs to buy for the waiting room. Figure 14.9 shows a possible description of this situation in Petri net form.

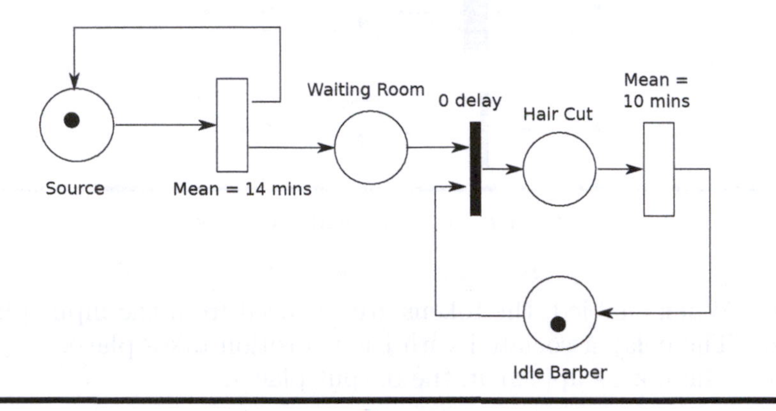

Figure 14.9 A barber's shop.

The source provides a new token to the waiting room in accordance with a Poisson process with mean interarrival time of 14 minutes. The source itself is also replenished by the transition and so never runs out of customers.

Once there is at least one customer in the waiting room and an idle barber, a haircut immediately starts (0 delay transition). The haircut takes a mean time of 10 minutes, and then the barber once again becomes idle, the token representing the customer having disappeared.

If this Petri net is fed into an appropriate tool, then it can perform a simulation and determine the mean number of people in the waiting room and the percentage of time for which the barber's chair is occupied (i.e., the barber is busy); see Figure 14.10. This figure has been produced by using the open-source PIPE (Platform Independent Petri Net Editor) tool and it can be seen, for example, that the barber is busy (there's someone in the barber's chair) for 57.9% of the time with a 95% confidence interval of ±1.3%.

Embedded software engineers are not often called upon to estimate the waiting area needed in a barber's shop. The relevance of the principle to messages arriving at a processor and needing to be stored before being processed is, I hope, clear.

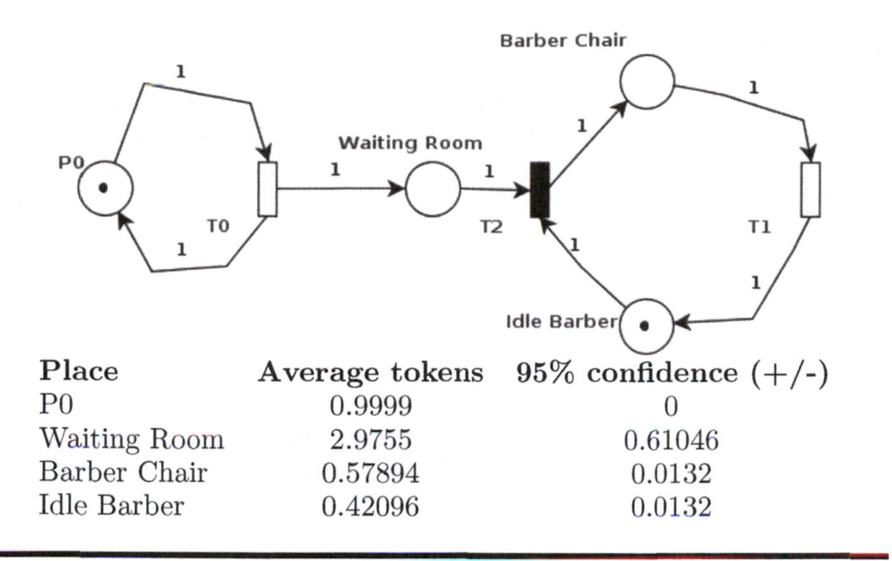

Place	Average tokens	95% confidence (+/-)
P0	0.9999	0
Waiting Room	2.9755	0.61046
Barber Chair	0.57894	0.0132
Idle Barber	0.42096	0.0132

Figure 14.10 Barber's shop: PIPE simulation output.

Petri Nets and Markov Models

One particularly easy transformation is creating a Petri net from a Markov model of the form shown in Figure 11.2 on page 159. Each state becomes a Petri net place, and a negative exponential transition is inserted on each arrow. A single token represents the current system state and is placed initially in State 1.

The Petri net tool will then be able to calculate all of the results given in Table 11.1 on page 161.

While this is a useful way of expressing a Markov model, having all transitions as negatively exponential in this way does not exploit the full flexibility of Petri nets.

What Do the Tokens Represent?

Tokens and places can represent anything.

When the Petri net is being used to represent a Markov model, there will be one place for each state and one token representing the current state.

In Figure 14.9, the places are more abstract. The place representing the waiting room is perhaps the least abstract, but the place representing the source is completely abstract. In this case, the tokens represent the barber and the customers.

Colored Petri Nets

The simple Petri net in Figure 14.9 can easily be extended. If there were two types of customer for the barber, perhaps adults and teenagers, each taking a different time for a haircut, then two sources could be provided, both feeding into the same waiting room. However, to handle the two types differently, the tokens would need to be distinguishable.

This distinction is provided by the concept of "colored" tokens: tokens that carry information with them. In this case, the information would be as simple as a type ("adult" or "teenager"), but it could be more complex.

A transition could then be programmed to fire only if there were, for example, 3 tokens of one "color" and 2 tokens of another "color" in its input places.

Petri Nets and Non-Determinism

What happens when two transitions are simultaneously enabled by a token arriving in a place?

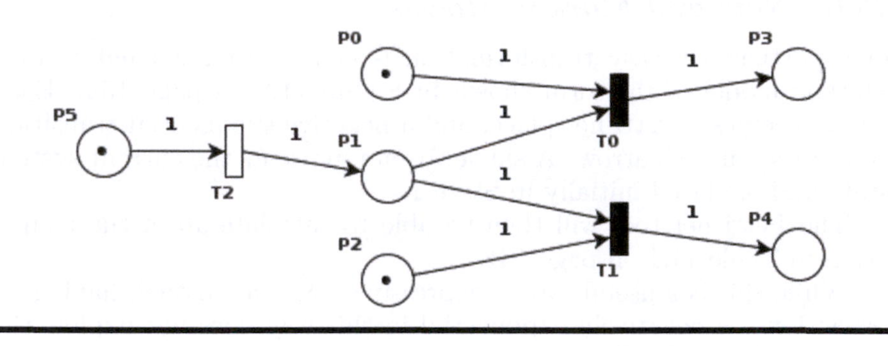

Figure 14.11 Non-determinism in a Petri net.

As an example, consider the network in Figure 14.11. Neither T0 nor T1 can fire because of the lack of a token in P1. If, however, T2 fires, then a token will appear in P1 and both T0 and T1 will be able to fire. Whichever one then fires first will remove the token from P1, thereby disabling the other.

Once T2 has fired, the behavior of this network is non-deterministic; it may exhibit different behavior on different runs. Rather than being a drawback in the modeling language, this actually reflects conditions of concurrency intrinsic to the system being modeled, and therefore essential to the model. In the system being modeled, two interrupts might occur at almost exactly the same time and be handled in a different order at different times. The modeling language should be powerful enough to allow this type of concurrency to be modeled so that the correctness of the design may be demonstrated under all conditions.

In Chapter 15, a description of the Promela language is given: another example of a non-deterministic language.

Petri Net Idioms

The Petri net is a language and, as with any language, there are idioms which, when known, make the language easier to read and write. Reference [7] by Winfrid G. Schneeweiss is a useful picture book containing many of these idioms.

Schneeweiss' reference [8] contains clichés relevant to applying Petri nets to the modeling of system dependability.

Simulation and the Example Companies

The representative companies introduced in Chapter 4 would be expected to deploy simulation, whether using a discrete event simulation tool or Petri nets, in different ways.

Alpha Device Corporation (ADC) makes complete devices, including both hardware and software, and will need some evidence that the software continues to operate correctly in the face of any combination and timing of inputs from the hardware. If possible, some of this evidence, in particular that related to liveness conditions (see page 18 for the distinction between safety and liveness conditions), would be provided by formally proving its correctness (Chapter 15). However, the evidence to support timing and availability claims will probably come from simulation — discrete event simulation or Petri nets.

If the device being created by ADC has a particularly complex user interface, another possible use of simulation is to allow potential customers to experiment with that interface before the design is complete. This allows potential user interface problems to be identified before the product is shipped. Reference [9] by Harold Thimbleby includes an interesting simulation of medical professionals making mistakes on different types of keyboard, while trying to enter a number, primarily to judge which type of keyboard is likely to result in the fewest mistakes.

Beta Component Incorporated (BCI), on the other hand, is building a software-only product that will be a component in ADC's device.

Simulation in their case would provide useful evidence that sufficient resources — memory, file descriptors, message blocks, etc. — had been allocated to ensure that the system would not fail because of an unavailable resource, particularly as resources are normally allocated statically in safety-critical systems — as recommended in IEC 61508-3, Table A.2.

Anecdote 19 *Some years ago, the company for which I was then working was designing a device with a complex user interface for a large customer. I built a formal model of the device, and the customer was invited to spend a day with us experimenting with the user interface. During the exploration, many problems were found with the interface, and I fixed these in the model as the trial progressed. At the end of the trial, the customer asked to sign a printed copy of the final model as the contract between us because it now unambiguously defined how the device should operate.*

I was delighted that my model had been such a success — I had invented the language and written the tool myself, but it soon became clear that I was a naïf engineer. My company refused to accept the specification.

In my politest manner, I asked why, and was told that an unambiguous specification left no room for wriggling when the product was finally delivered. Instead, my company insisted on using an ambiguous, English-language specification.

The moral of the story is that, even though engineers want unambiguous requirements, and the standards encourage this, it may not always be commercially acceptable.

References

1. D. Ganesan, M. Lindvall, R. Cleaveland, R. Jetley, P. Jones, and Y. Zhang, "Architecture reconstruction and analysis of medical device software," in *Proceedings of the 2011 Ninth Working IEEE/IFIP Conference on Software Architecture*, WICSA '11, (Washington, DC, USA), pp. 194–203, IEEE Computer Society, 2011.
2. T. J. Schriber, *Simulation using GPSS*. John Wiley and Sons, 1974.

3. J. von Neumann, "Various Techniques Used in Connection with Random Digits," *J. Res. Nat. Bur. Stand.*, vol. 12, pp. 36–38, 1951.
4. D. E. Knuth, *The Art of Computer Programming, Volume 2 (3rd ed.): Seminumerical Algorithms.* Boston, MA, USA: Addison-Wesley Longman Publishing Co., Inc., 1997.
5. C. A. Petri, *Kommunikation mit Automaten.* PhD thesis, Institut für instrumentelle Mathematik, Bonn, 1962.
6. M. A. Marsan, "Stochastic Petri Nets: An Elementary Introduction," *Lecture Notes in Computer Science; Advances in Petri Nets 1989*, vol. 424, pp. 1–29, 1990. NewsletterInfo: 36.
7. W. Schneeweiss, *Petri Net Picture Book.* LiLoLe-Verlag GmbH, 2004.
8. W. Schneeweiss, *Petri Nets for Reliability Modeling.* LiLoLe-Verlag GmbH, 1999.
9. H. Thimbleby, "Safety versus Security in Healthcare IT," in *Addressing Systems Safety Challenges, Proceedings of the 22nd Safety-Critical Systems Symposium* (C. Dale and T. Anderson, eds.), pp. 133–146, 2014.

Chapter 15

Formal Design Verification

Formal methods should be part of the education of every computer scientist and software engineer, just as the appropriate branch of applied mathematics is a necessary part of the education of all other engineers. Formal methods provide the intellectual underpinnings of our field; they can shape our thinking and help direct our approach to problems along productive paths; they provide notations for documenting requirements and designs, and for communicating our thoughts to others in a precise and perspicuous manner; and they provide us with analytical tools for calculating the properties and consequences of the requirements and designs that we document.

John Rushby in reference [1]

Chapter 14 covered some of the semi-formal techniques that can be used to answer the question: "Will a system implemented to this design deliver what is required of it?" This chapter addresses the same question, but using formal methods.

What Are Formal Methods?

Page 17 provides an introduction to the distinction between formal, semi-formal, and informal languages. A design can be expressed informally (for example, in English), semi-formally (for example, as a Petri net), or formally. The formal language is mathematics, although the tools used to create designs expressed in a formal language hide some, if not all, of the mathematics from the user.

Once the design of a system, a component, or an algorithm is expressed formally, we can reason about it. Will the algorithm always complete within a finite time irrespective of the combination and timing of events that occur? Does the receipt of message A *always* lead to message B being sent? Is there *any* combination of circumstances under which message B would be sent without message A having been previously received? Can the processes deadlock? If process A attempts to send message X infinitely often, then will it succeed infinitely often?

However, a formal design can provide more than verification. It should be possible to generate the requisite software automatically from a validated formal design. In principle, it should not be necessary to test the implementation; if code has been produced by a trusted algorithm from a design proven to be correct, why would testing be necessary? In practice, testing will be required, but as described below, IEC 61508 accepts that, when formal methods are used, it is expected that the testing activity will be reduced. Remember Donald Knuth's 1977 warning: "Beware of bugs in the above code; I have only proved it correct, not tried it."

History of Formal Methods

Perhaps the first practical formal language for expressing system design was the Vienna Development Method (VDM) originating from IBM's Vienna laboratory in the 1970s. This work was published in 1978 by Dines Bjørner and Cliff B. Jones in reference [2].

In parallel with the VDM, Jean-Raymond Abrial was working on the Z language, which also appeared in print in the late 1970s. I describe the Rodin tool below and this operates with the Event-B language, which is a derivative of the B language (also constructed by Abrial), which is itself a direct descendent of Z.

These languages rely on set theory, the lambda calculus, and first-order predicate logic, although what is demanded of the user is little more than a basic knowledge of set theory, including functions, and a willingness to learn the mathematical notation.

At the same time as the Z/B/Event-B development, the Spin tool was developed in the Computing Sciences Research Centre at Bell Labs by Gerard J. Holzmann and others, beginning in 1980. Rather than set theory, Spin is based on non-deterministic Büchi automata, named after Julius Richard Büchi, a Swiss mathematician of the middle 20th century. Again, almost all of this complexity is hidden from the user.

Since those early days, these languages, and many others, have been enhanced, particularly in the areas of proving algorithms, and practical tools have become available.

Formal Methods and the Standards

IEC 61508

Paragraphs 7.4.7 and 7.4.8 of IEC 61508-3 accept that when formal methods are used, the number of test cases required for verification may be reduced. Tables A.1, A.2, and A.4 recommend the use of formal methods for the specification of the safety requirements and for specifying the software design. The use of formal methods for testing is recommended in Tables A.5 and A.9. Additionally, Table A.7 calls up Table B.5 which recommends formal methods as a modeling tool.

Paragraph B.2.2 of IEC 61508-7 describes formal methods and specifies the aim of using them, as follows:

> *Formal methods transfer the principles of mathematical reasoning to the specification and implementation of technical systems [thereby increasing] the completeness, consistency or correctness of a specification or implementation.*

That paragraph of IEC 61508 goes on to list some of the disadvantages of formal methods. These are points to avoid when using them for a product that needs to be approved to IEC 61508:*

IEC 61508: "Fixed level of abstraction."
: The Rodin tool described below avoids this by having a hierarchy of abstractions. At the lowest level, the specification is completely abstract. At each step, the specification becomes more concrete, until it is finally completely concrete and code can be generated.

IEC 61508: "Limitations to capture all functionality that is relevant at the given stage."
: It is not clear what is meant here; certainly the person building the model has to ensure that all relevant functionality has been

* I have retained the precise wording from IEC 61508, even when the English appears to be somewhat strained.

included, and designers are sometimes unable to model all aspects of the system formally. As reference [3] by P.C.Ölveczky and J. Meseguer states:

> *At present, designers of real-time systems face a dilemma between expressiveness and automatic verification. If they can specify some aspects of their system in a more restricted automaton-based formalism, then automatic verification of system properties may be obtained by specialized model checking decision procedures. But this may be difficult or impossible for more complex system components. ... In that case, simulation offers greater modeling flexibility, but is typically quite weak in the kinds of formal analyses that can be performed.*

The real-time Maude language has specifically been designed to bridge the gap between formality and expressiveness.

IEC 61508: "Difficulty that implementation engineers have to understand the model."

This can be a problem. As pointed out by the present author, Guy Broadfoot, and Akramul Azim in reference [4], this is not only a problem for implementation engineers. Few auditors, confronted with pages of theorems and proofs supposedly justifying the correctness of a design are able to absorb them and assess their correctness.

IEC 61508: "High efforts to develop, analyze and maintain model over the lifecycle of system."

There *is* a project cost associated with developing formal models but, as IEC 61508 points out, this can be offset somewhat by the reduction in testing time. Automatic code generation from the formal model can also reduce the total project time — see page 218.

I believe that it is probable (see, for example, reference [5] by Steve King *et al*) that the economic balance is tipping away from verification through testing to verification by means of formal methods.

IEC 61508: "Availability of efficient tools which support the building and analysis of model."

This is something that has been improving over the 21st century as computer power and algorithms have improved. The Spin tool described below has been around since 1980, but the increase in computing power and memory has made it much more useful recently. The Rodin tool, also described below, was created in the first decade of the 21st century.

IEC 61508: "Availability of staff capable to develop and analyze model."
It is true that it is difficult to find staff capable of developing formal models. As the quotation at the head of this chapter says, more education is needed, but, as shown in the examples below, the level of mathematical maturity required is not great.

I feel that the core question is, who should be producing the models? Development teams traditionally consist of people who are programmers at heart — they may be working as designers, implementers, or testers, but they are really programmers. The creation of a formal model using Spin's Promela language as described below is very close to writing a C program. This can be very confusing because, in spite of the syntactic similarity, a Promela program is *not* a computer program, it is a definition of a design. It would perhaps be better not to use programmers for producing models because, although the two skills are related, they are actually distinct.

IEC 61508: "The formal methods community's focus clearly [has] been the modeling of the target function of system often deemphasizing the fault robustness of a system."
This is a question of how the tools are applied: robustness and functionality can be modeled. Ken Robinson of the University of New South Wales has pointed out on several occasions that Event-B can certainly handle many nonfunctional requirements including capacity, timing, and availability.

ISO 26262

ISO 26262-6 recommends formal methods for various activities. Tables 2 and 7 recommend using a formal notation for capturing a software architecture and unit design, and Tables 6 and 9 recommend formal verification for systems that need to conform to automotive safety integrity level (ASIL) C or D.

IEC 62304

IEC 62304's only reference to formal methods is somewhat negative. Paragraph 5.2.6 states, "This standard does not require the use of a formal specification language."

Do Formal Methods Work?

As long ago as 1990, Anthony Hall wrote a paper (reference [6]) identifying some myths associated with formal methods. Here are the myths he lists, together with my contemporary comments on them:

Myth: Formal methods can guarantee that software is perfect.
Actually, they only work on models of the system; it is the analyst's job to ensure that the model sufficiently reflects reality.

Myth: Formal methods are all about program proving.
Proving that programs were correct was a particular fashion in the late 1980s and early 1990s when Hall's paper was written. Formal methods as described here are not for reasoning about programs; they are for reasoning about models of systems.

Myth: Formal methods are only useful for safety-critical systems.
They are also useful for security- and mission-critical systems — any system where dependability is required.

Myth: Formal methods require highly-trained mathematicians.
Mathematical notation and some mathematical thinking is involved, but highly trained mathematicians are not required, most of the mathematics being quite simple and hidden by the tools.

Myth: Formal methods increase the cost of development.
This is particularly pernicious. Actually, their use brings cost forward in the project, increasing the cost of architectural and design work and decreasing the cost of implementation and testing. However, managers often assess the progress of a project by the progress of the implementation (coding), and this gives rise to this myth.

Myth: Formal methods are unacceptable to users.
See Anecdote 19 on page 208. As that anecdote shows, the customer (the user) found the formal expression of their requirements very acceptable.

Myth: Formal methods are not used on real, large-scale software.
There are several highly successful projects quoted in analyses of formal methodologies, e.g., the Paris Metro and the Rotterdam Flood Barrier. The frequency with which the same examples occur in the literature may be a reflection that there is some truth in this myth, although other examples are beginning to emerge. For example, reference [7] by Chris Newcombe *et al.* describes how formal methods, using TLA$^+$, are applied to distributed algorithms by Amazon Web Services. The three key insights in that study are as follows:

1. Formal methods find bugs in system designs that cannot be found through any other technique we know of.
2. Formal methods are surprisingly feasible for mainstream software development and give good return on investment.
3. At Amazon, formal methods are routinely applied to the design of complex real-world software, including public cloud services

Reference [4] refers to the successful application of formal methods (using the Rodin tool; see page 225) to the definition of a medical device.

Types of Formal Methods

It is difficult to classify cleanly the types of formal languages and tools that are available, but the following may provide a useful taxonomy:

Model-oriented techniques.
: With these the analyst defines the system in terms of its behavior: "When X occurs the system does Y." Z, B, Event-B, TLA^+ and VDM are examples of languages built around this approach.

Property-oriented techniques.
: Here, the analyst concentrates on properties rather than the behavior of the system: "It is always true that system state X eventually occurs." Larch, OBJ, CASL, Clear, and ACT One are examples of this approach. Linear Temporal Logic also fits here; see page 330 and its application in Spin on page 223.

Process algebra techniques.
: These are largely based on C.A.R. Hoare's original (1978) work on Communicating Sequential Processes: reference [8]. They deal with the complex interactions between different processes in a system.

In the following sections, I consider two tools in more detail: Spin and Rodin. I have chosen these from the wide range of possible languages and tools because they are very different from each other, because they are the tools with which I am most familiar, and because they are available as open-source meaning that the reader can experiment with them at no cost and can, if necessary, extend the tool.

Automatic Code Generation

The intention of formal design verification is to produce a mathematically complete definition of the design and then prove that the design is correct according to chosen invariants that must remain true irrespective of the sequence of events to which the system is subjected.

Given such a complete definition of the design, it should be possible to generate the code to realize the design automatically. For some tools this is the case: the Rodin tool described on page 225 has a code generator for producing Java, C, C++, and C# code. Commercial tools, such as VerumST's Dezyne, create high-quality code.

Automatic code generation is a mixed blessing: it requires a change in the structure of development teams — perhaps 5 designers and 3 programmers rather than 2 designers and 10 programmers — and it may complicate the approvals process when the quality of the code generator will need to be justified.

It also complicates code maintenance. It is unlikely that the code has been generated in accordance with the company's coding standards* and a decision has to be made about whether the generated code should ever be modified by a human, or whether the model should always be changed and the code regenerated.

One major advantage that could accrue from this form of automated code generation, but which does not seem to be available from the existing tools, is the parallel generation of good integration test cases. Reference [9] by the current author outlines some of the dangers that we face by generating module tests from the code (to achieve the code coverage metrics demanded, for example, by IEC 61508). Generating module and integration tests from the formal system design avoids that danger and should provide high-quality test cases.

Spin Modeling Tool

Spin's History

Spin is a mature tool whose origins can be traced back to the 1980s, and it can be used in simulation and verification modes to debug a design and then prove its correctness (or find a counter-example). During verification, it will try to prove the truth of invariants specified using assertions or linear temporal logic (LTL); see page 330.

* It is interesting that some commercial tools, for example, ETAS's AutoSAR OS generation tool, not only produce code in accordance with the MISRA C standards, but also generate a report of the level of compliance and justification for violations.

Spin is released under a variety of licences, including GPLv2.

Spin's Structure

This section can only touch the surface of the Spin modeling system. For more details refer to the classical text, Gerard Holzmann's book, reference [10].

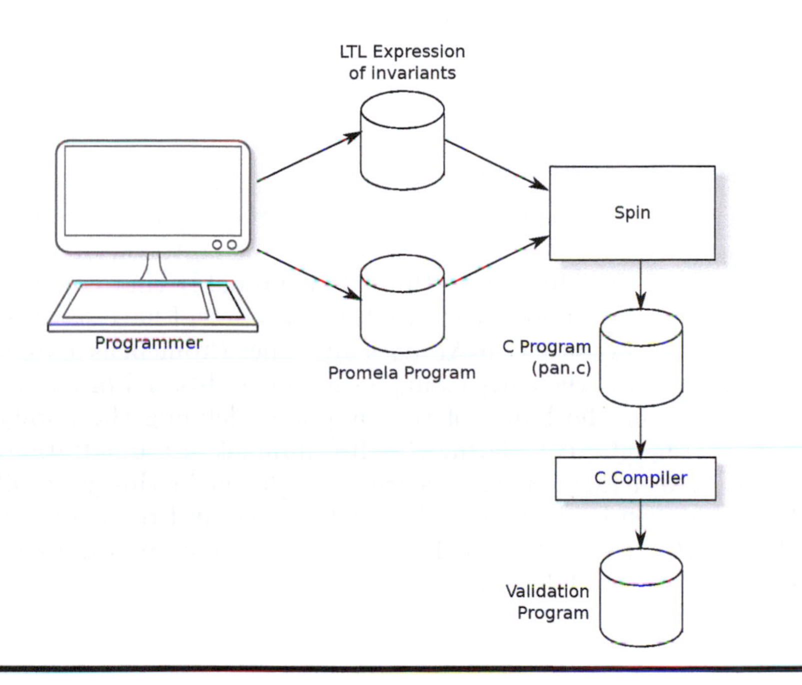

Figure 15.1 Using Spin.

Figure 15.1 illustrates the process used to create and verify a Spin model:

The programmer creates a model of the system under investigation using a language known as Promela (a contraction of PROcess MEtaLAnguage — see below).

If there are invariants that need to be expressed in linear temporal logic (LTL — see page 330), then the programmer codes these using a keyboard-friendly form of LTL. For example, $\diamond$ ("eventually") becomes <> and $\Box$ ("always") becomes [ ].

Spin converts the linear temporal logic statements into Promela and combines these with the Promela that the programmer produced to create a C program that will carry out the verification.

The C program is compiled and executed in the normal way. The result is either a proof that the invariants are always true, or a sequence of events that results in an invariant being violated because of a fault in the design.

These and related steps (e.g., simulating the system to check whether it has been correctly specified) can be carried out manually using Spin's command-line options, or can be hidden behind a graphical user interface, such as `ispin` or `jspin`.

The Promela Language

Promela is not a programming language; rather, it is a language for describing the design of a system, algorithm, or protocol.

The gentlest introduction to Promela is probably that contained in Mordechai Ben-Ari's book, reference [11]. However, I have reservations about that book because Ben-Ari *does* introduce Promela as a programming language, and the early examples are of traditional programming exercises: finding the larger of two numbers, defining the number of days in each month, calculating the discriminant of a quadratic equation, etc. I feel that someone learning Promela in this way will be confused between Promela and C, on whose syntax Promela is based.

One characteristic of Promela is that it is a non-deterministic language. Consider the following piece of C code:

```
if (x == 2) y = 32;
if (x < 7) y = 18;
```

If this routine is entered with $x = 2$, then both conditions will be found to be true and y will always end up as 18.

A similar piece of Promela code might read

```
if
    :: x == 2  ->  y = 32;
    :: x < 7  ->  y = 18
fi;
```

If this code is entered with x having a value of 2, then the two "guards" indicated by the double colons (::) are both true. Which of the two instructions are executed ($y = 32$ or $y = 18$) is non-deterministic.

This non-determinism is extremely useful in reflecting non-determinism in the system being modeled. Perhaps within the system interrupt A may occur slightly before or slightly after interrupt B; either event can occur. The snippet of Promela code above is similar, stating that, if x is 2, then y may become either 32 or 18.

There are some other syntactic differences between Promela and C, for example the semicolon (;) is a separator in Promela, but a terminator in C.

Using Spin to Check a Mutual Exclusion Algorithm

To illustrate the use of Spin in an application so simple that LTL is not required, I will take a mutual exclusion algorithm published in 1966. As the algorithm contains an error, I will not provide a reference to the paper, but I will say that the journal in which it was published was, and is, prestigious.

The purpose of a mutual exclusion algorithm is to provide a thread, Thread A, with access to a particular resource, other threads being excluded until Thread A releases it. In the case of the published algorithm, there are two threads competing for the resource: Thread 0 and Thread 1.

Three variables are used to allow the two threads to avoid clashing over the resource:

1. `flag[0]`, which is initialized to 0
2. `flag[1]`, which is initialized to 0
3. `turn`, which is initialized to 0 or 1

When Thread i wants to gain exclusive access to the resource, then it carries out the following algorithm:

1. Set `flag[i]` to 1
2. While `turn` is not i:
   - Wait for `flag[1 - i]` to become 0
   - Set `turn` to i
3. Access the resource, the other thread being excluded
4. Set `flag[i]` to 0

The PROMELA code corresponding to this algorithm is as shown in Figure 15.2. Note that to ensure that both threads are never in the critical section at the same time, each one increments and then decrements the variable `critical`. If the algorithm is correct, then this variable can never be greater than 1 — hence the `assert()` statement.

```
   bit turn = 0;
   bit flag[2];
   byte critical = 0;

 1 proctype P(byte i)
 2     {
 3     flag[i] = 1;
 4
 5     do
 6         :: turn != i ->
 7             if
 8                 :: flag[1 - i] == 0 -> turn = i
 9             fi;
10
11         :: else -> break
12     od;
13
14     critical++;  /* in the critical section */
15
16     /* check only one thread in critical section */
17     assert(critical < 2);
18
19     critical--;   /* leaving the critical section */
20     flag[i] = 0;
21     }
22
23 init {
24     flag[0] = 0;
25     flag[1] = 0;
26     atomic {
27         run P(0);
28         run P(1);
29         }
30     }
```

Figure 15.2 Promela code for mutual exclusion algorithm.

When executed, Spin immediately finds a series of events that lead to both threads being in their critical sections at the same time, indicated by `critical` being greater than 1. The sequence of events is:

1. Thread 1 needs to access the resource and sets `flag[1]` = 1
2. Thread 1 finds that `turn` is not 1 (it was initialized to 0)
3. Thread 1 finds `flag[0]` is 0
4. At this point, thread 0 wakes up and needs to access the resource. It sets `flag[0]` = 1
5. Thread 0 finds that `turn` is 0 and so enters its critical section
6. While Thread 0 is in its critical section, Thread 1 sets `turn` = 1
7. Thread 1 can now enter its critical section

Both threads now believe that they have exclusive access to the shared resource! In terms of the model, each increments `critical` and, when the second one does so, the `assert()` statement detects the error condition.

We now have the necessary tools to prove that the program given in Figure 5.4 on page 70 contains a serious fault. Converting this program to Promela is straightforward, and an `assert(x != 5)` statement immediately before the `printf()` quickly produces the sequence of events that will lead to a division by zero.

Spin and Linear Temporal Logic

Neither the simple mutual exclusion algorithm nor the program in Figure 5.4 actually needs the use of linear temporal logic (LTL), but both can make use of it because in both cases we are using the `assert()` statement to check that some condition *never* occurs. In the case of the mutual exclusion algorithm, we are interested in proving the value of `critical` can *never* be greater than 1; in the case of Figure 5.4 we are interested in the condition that x can *never* be 5 after the two threads have completed their execution.

Using the LTL notation described on page 330, the condition that `critical` can never be greater than 1 can be expressed as:

#define exclusion (critical <= 1)
$\Box$ exclusion

where $\Box$ is the symbol meaning "it is always true that ..."

Spin can be used to turn the negation of such an LTL statement into Promela by using the `-f` command-line option as illustrated in Figure 15.3. Note that this produces a `never` clause, which leads to some twisted thinking with double negatives. The code checks the claim that it is *not* ($\neg$ or ! in ASCII) true that `exclusion` is always true ($\Box$ exclusion).

This generated Promela code can be copied into the program, and

```
$ spin -f '![]exclusion'

never {    /* ![]exclusion */
T0_init:
do
:: atomic { (! ((exclusion))) -> assert(!(! ((exclusion)))) }
:: (1) -> goto T0_init
od;
accept_all:
skip
}
```

Figure 15.3 Spin creating a "never" clause.

the `assert()` statement on line 17 can be removed. Using the LTL is safer than putting `assert()` statements into the code manually, because the LTL statement will be checked after every Promela statement, not just at the point where we have manually placed the `assert()`. Running the Spin checker with this additional Promela gives the same result.

Using the terminology described on page 18, this LTL is checking a "safety" property. It is also possible to generate the Promela code to check "liveness" properties. For example, the negation of the property $\Diamond\Box P$ can be checked by the Promela code

```
$ spin -f '!<>[]p'
never {    /* !<>[]p */
T0_init:
do
:: (! ((p))) -> goto accept_S9
:: (1) -> goto T0_init
od;
accept_S9:
do
:: (1) -> goto T0_init
od;
}
```

Spin Summary

This section has given a very superficial description of the Spin tool (reference [10] has a full description and is 596 pages long!).

In practice, although the algorithms used within Spin are efficient, there is a maximum size of the system's state space that Spin can handle. Reference [10] contains a number of hints for reducing the state space of a model, but ultimately, the size of the model may become too large for Spin to handle. I have found Spin to be really useful for demonstrating the correctness of algorithmic parts of a design, rather than a complete design.

Rodin Modeling Tool

Rodin's History

Rodin development started in 2004 within the Rodin project funded by the European Commission. The funding was continued in the Deploy project until 2012 and then in the Advance project until 2014. Various industrial partners supported the development, including Bosch, Siemens Transportation Systems, and Space Systems Finland.

At the time of writing, the Rodin development is still very active, with frequent new releases and much discussion on the associated websites.

Rodin is based on the Eclipse framework and is released under the Eclipse Public License v1.0.

Rodin's Structure

Figure 15.4 illustrates the concept underlying Rodin. Initially, the analyst defines a very abstract model of the system in three parts:

A context.
: This defines the static information about the model at a particular level of abstraction — constants sets, etc.

A machine.
: This describes the state changes that can occur in the model at that level of abstraction.

A list of invariants.
: These are conditions that must be unconditionally true in the model at that level of abstraction and all the less abstract levels.

The contexts and machines are then progressively made more concrete by refining algorithms, interactions, and events. For example, an abstract algorithm may simply state that "an input produces an acceptable output" without describing how that is achieved. A refinement would then add the details of *how* the input is transformed into the

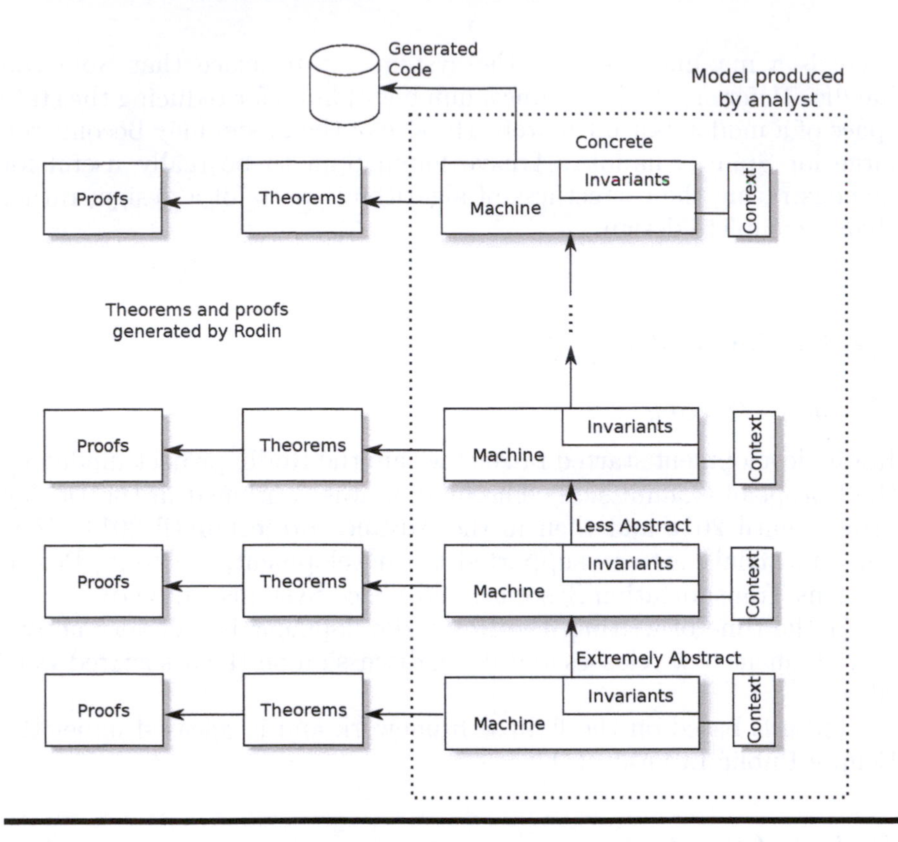

Figure 15.4 The Rodin concept.

output. Similarly, an abstract interaction of the type "an approved operator may start the flow of drugs to the patient" could be made more concrete at the next level by indicating how the operator becomes "approved" — perhaps by entering a password. Events can also be refined. So the abstract event "patient needs attention" might be refined to "the patient's heart rate goes outside the range 45 to 100."

At each level of refinement, not only must the new invariants added at that level be universally true, so must all the invariants defined at more abstract levels.

Ultimately, when the model is completely concrete and has been proven to be correct, it should be possible to generate code automatically from it, and various plug-ins exists to do just that.

Rodin provides assistance during this process. Having checked the syntax of the context, machine, and invariants, Rodin automatically generates the theorems that will need to be proven to ensure that the model is well-formed and that the invariants are always true.

Having generated the theorems, Rodin uses one or more of its theorem-prover plug-ins (e.g., proB) to attempt to prove them. If the theorem prover cannot find a proof without help, the analyst can offer lemmata: effectively, suggestions on how the proof might be generated. Rodin does not accept these at face value; instead, it adds them to its list of things to be proven, but providing lemmata can move the theorem prover into a new direction that leads to a complete proof.

Being Eclipse-based, Rodin has numerous plug-ins available to enhance the base product. These include a powerful LTL engine (see page 330) that can check for acceptance cycles, a form of liveness check where an invariant can be written to express the condition that a particular part of the design is always executed infinitely often. Other plug-ins include a B2LaTeX program that will produce a LaTeX document from the Rodin model and the ProR requirements capture tool.

Anecdote 20 *The automatic theorem provers that come with Rodin are very powerful. With one exception, when I have spent time shouting at the computer and tearing my hair out trying to see why the stupid thing couldn't prove what to me was an obviously correct theorem, I have found that I have overlooked a corner case and, in fact, that the theorem was false — the tool was correct and I was wrong. In such a case, it is best to apologize to the computer for shouting at it, to avoid problems with the next theorem.*

In spite of the existence of B2LaTeX it is probably worth saying that I find one of Rodin's main weaknesses to be the impossibility of generating good summary reports about the model definition, the theorems, and proofs.

Using Rodin

Rodin is a tool for manipulating the Event-B language and is based on the Eclipse framework. Anyone familiar with Eclipse will immediately feel at home with Rodin; see Figure 15.5 which illustrates the screen used for editing contexts and machines.

Area A lists the various projects, and area B is where the editing takes place. Because non-ASCII mathematical symbols need to be used in contexts and models, area C provides a palette of these, although they can all be typed using combinations of keyboard keys (for example,

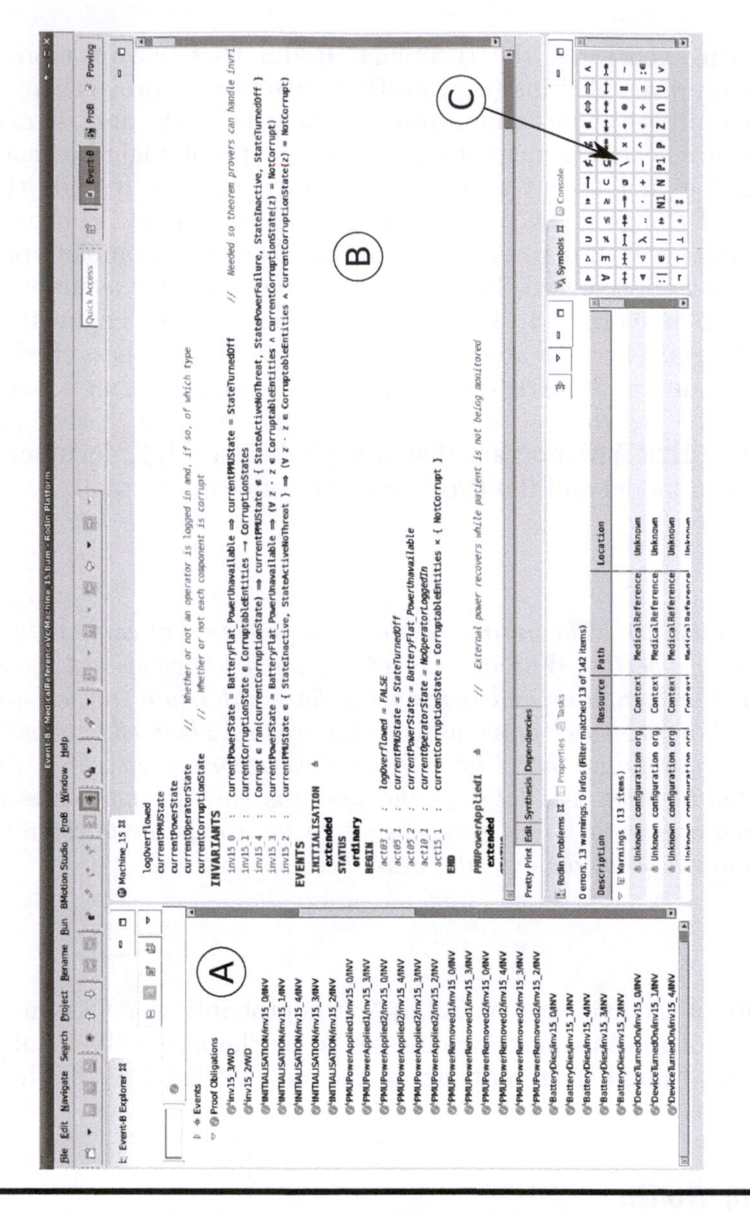

Figure 15.5 The Rodin concept.

∃ can be typed as "#" and ∈ as ":"). Ken Robinson has published a very useful "cheat sheet" of the Rodin operators and how to type them.*

Rodin includes various editors for creating contexts and machines.

* Available at `http://handbook.event-b.org/current/html/files/EventB-Summary.pdf`.

In particular, the Camille plug-in seems to be very popular, but I prefer the rather clunky built-in editor, perhaps because that's the one with which I started using Rodin.

Rodin's Theorem Proving

Once one or more layers of the model have been created, Rodin generates the theorems necessary to ensure that invariants cannot be broken. Its plug-ins then attempt to prove those theorems; see Figure 15.6.

For each theorem, the *proving perspective** contains the goal that it is trying to prove (window B), the proof steps so far (window A), the step on which it is currently working (window D), and, most important of all, a smiling or miserable face (point C) indicating whether the proof attempt has been successful or not. In Figure 15.6, the face is smiling, indicating that a proof has been found.

Checking the Mutual Exclusion Algorithm Again

Applying Rodin to the mutual exclusion model described on page 221 is somewhat like using a pile driver to crack a nut, but I have applied it so that its output can be compared with that of Spin.

The key to the model is the invariant that the two processes must never be in their critical sections at the same time. Rather than use the trick of incrementing and decrementing a counter as I did with the Spin model, I code the following invariant into the Rodin model:

$$\neg(\text{proc0State} = \text{ProcessCritical} \wedge \text{proc1State} = \text{ProcessCritical})$$

That is, it is not ($\neg$) true that process 0's state and ($\wedge$) process 1's state are both critical at the same time.

It takes Rodin almost no time to find that this invariant can be violated; see Figure 15.7 where the violation is identified (label A) and the series of events leading to that violation are listed (label B; read the events from the bottom to the top).

It is interesting that Rodin has found a slightly different sequence of events from that found by Spin.

As part of the initialization, I specified that

$$\text{turn} \in \{0, 1\}$$

i.e., that `turn` could start with a value of either 0 or 1. Rodin's path

* In Eclipse terminology, each type of operation has its own perspective, i.e., collection of windows.

to the breaking of the invariant starts with $turn = 1$.

Note that, even for a model as simple as this, Rodin generated 17 theorems that had to be proven. It can be seen from Figure 15.8 that Rodin managed to find proofs for all except 2 invariants: invariants 6 and 8. These are the two invariants that say that both processes must not be in their critical sections simultaneously — clearly these theorems couldn't be proven, because they are not true!

Formal Modeling by the Example Companies

The representative companies introduced in chapter 4 will both probably make use of some form of formal verification of their designs.

Alpha Device Corporation (ADC) could reasonably apply Rodin to verify the overall design of its device and Beta Component Incorporated (BCI) could apply Spin to various low-level algorithms within the operating system kernel (scheduling algorithms, memory allocation algorithms, mutual exclusion algorithms between processor cores, etc.).

With this work completed, both ADC and BCI potentially have the problem pointed out on page 214 and reference [4]: How can these proofs be presented to the certification auditors?

Both Spin and Rodin are probably classified as TI2/TD2=TCL2 in the terminology used within section 11 of ISO 26262-8 and as T2 in the IEC 61508 terminology. Neither Spin nor Rodin directly affects the delivered code, but both can fail to reveal a defect that exists in the design. This means that not only must the output from the tools be presented to an auditor, the tools themselves will have to be justified.

The first of these is hard enough because, in the case of ADC's use of Rodin, a printed list of several hundred theorems and associated proofs may impress an auditor, but from my experience, many auditors will not be able to assess the value of those theorems and proofs. BCI has an even bigger hurdle to jump because Spin does not produce a report, meaning that the auditor will have to sit at the screen and use the `ispin` or `jspin` graphical interface.

Anecdote 21 *The fear of the mathematics is real. I once presented some of these formal methods to a group of programmers and one programmer said to me afterwards, "I went into programming so that I wouldn't have to do maths." He may have been joking.*

Another certification question is raised if ADC uses the model to generate code automatically. There are two possible types of argument associated with this code:

1. *The code could have been produced by a human; actually, it has been produced by a computer. Nothing in the verification phase depends on the source of the code — it will still be module tested, integration tested, and system tested and so its origin is irrelevant.*

   This argument is probably the more useful, but it forgets that if the code does not conform to the company's coding standard, it may be difficult to review. Moreover, it does not address the maintenance of the code — ADC needs to have a clearly defined process for how the code will be modified if changes are needed: Will it be manually changed and, if so, will the coding standards then be applied?
2. *The code has been produced by a trusted algorithm and therefore does not need to be inspected or to have module testing performed on it.*

   Supporting this argument would mean reviewing the code generation algorithms, which could be tedious.

Formal Methods

Formal methods have promised much since at least the late 1980s, but have delivered relatively little. In part, this disappointing delivery has been due to technical limitations, but in larger part because the advocates of formal methods have not successfully bridged the gap between the theory and the working programmer.

I believe that improved algorithms and faster processors are reducing the technical limitations, and, to a large extent, tools such as Rodin are making formal methods accessible to the programmer at the keyboard. This does, however, raise the question that I mentioned earlier — who should be producing and verifying models of the design?

To avoid an impedance mismatch between design and implementation, design work is carried out in many organizations by the programmers who will have to implement it. I believe that the skills and tools needed for design and for programming are different, and there are few engineers who have mastery of both.

References

1. J. Rushby, "Formal Methods and the Certification of Critical Systems," Tech. Rep. SRI-CSL-93-7, Computer Science Laboratory, SRI International, Menlo Park, CA, Dec. 1993. Also issued under the title *Formal Methods and Digital Systems Validation for Airborne Systems* as NASA Contractor Report 4551, December 1993.
2. D. Bjørner and C. B. Jones, eds., *The Vienna Development Method: The Meta-Language*, vol. 61 of *Lecture Notes in Computer Science*, Springer, 1978.
3. P. C. Ölveczky and J. Meseguer, "Semantics and pragmatics of Real-Time Maude," *Higher-Order and Symbolic Computation*, vol. 20, no. 1-2, pp. 161–196, 2007.
4. C. Hobbs, G. Broadfoot, and A. Azim, "Hat das Zeitalter der formalen Methoden schon begonnen?," in *Proceedings of ESE Kongress 2012*, 2012.
5. S. King, J. Hammond, R. Chapman, and A. Pryor, "Is proof more cost-effective than testing?," *Software Engineering, IEEE Transactions on*, vol. 26, no. 8, pp. 675–686, 2000.
6. A. Hall, "Seven Myths of Formal Methods," *IEEE Softw.*, vol. 7, pp. 11–19, September 1990.
7. C. Newcombe, T. Rath, F. Zhang, B. Munteanu, M. Brooker, and M. Deardeuff, "How Amazon Web Services Uses Formal Methods," *Commun. ACM*, vol. 58, pp. 66–73, Mar. 2015.
8. C. A. R. Hoare, "Communicating Sequential Processes," *Commun. ACM*, vol. 21, pp. 666–677, Aug. 1978.
9. C. Hobbs, "Das Paradoxon dynamischer Softwaretests," in *Proceedings of ESE Kongress 2014*, 2014.
10. G. Holzmann, *The SPIN Model Checker: Primer and Reference Manual.* Addison-Wesley Professional, second ed., 2004.
11. M. Ben-Ari, *Principles of the Spin Model Checker.* Springer, 1 ed., 2008.

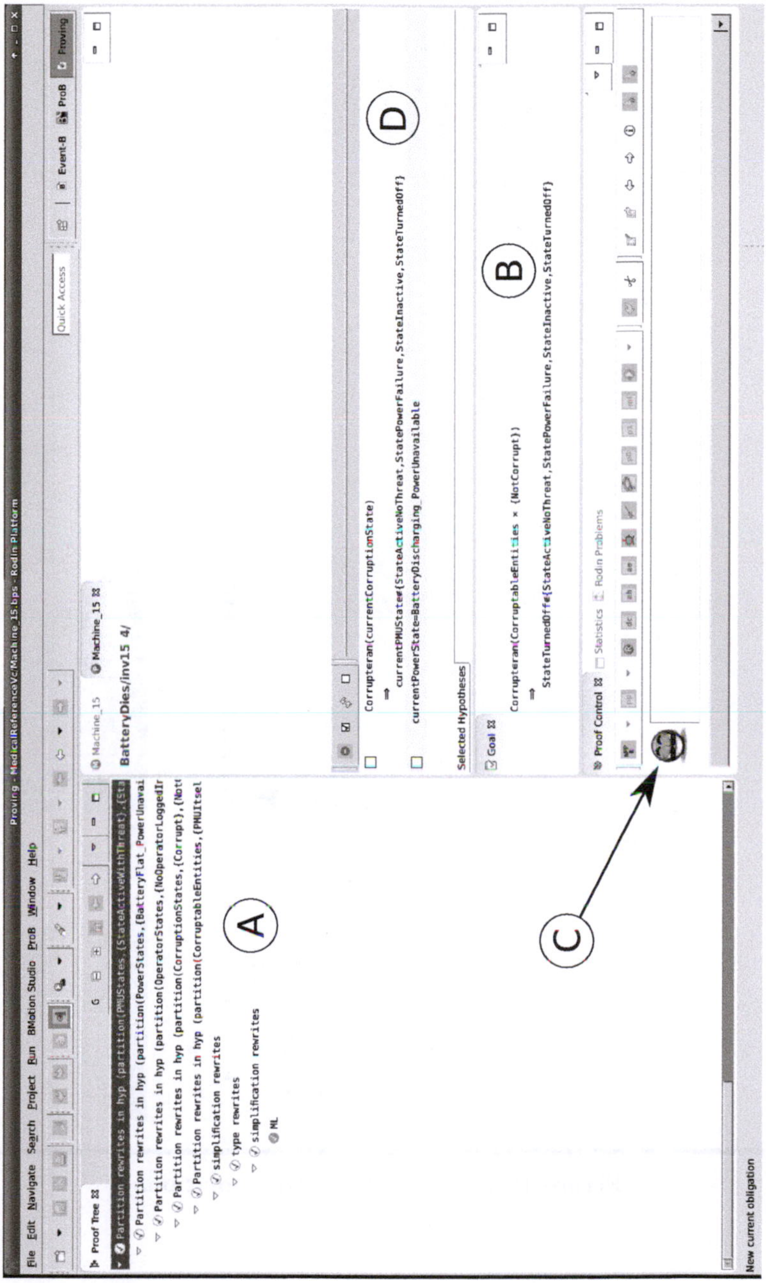

Figure 15.6 **The Rodin proving display.**

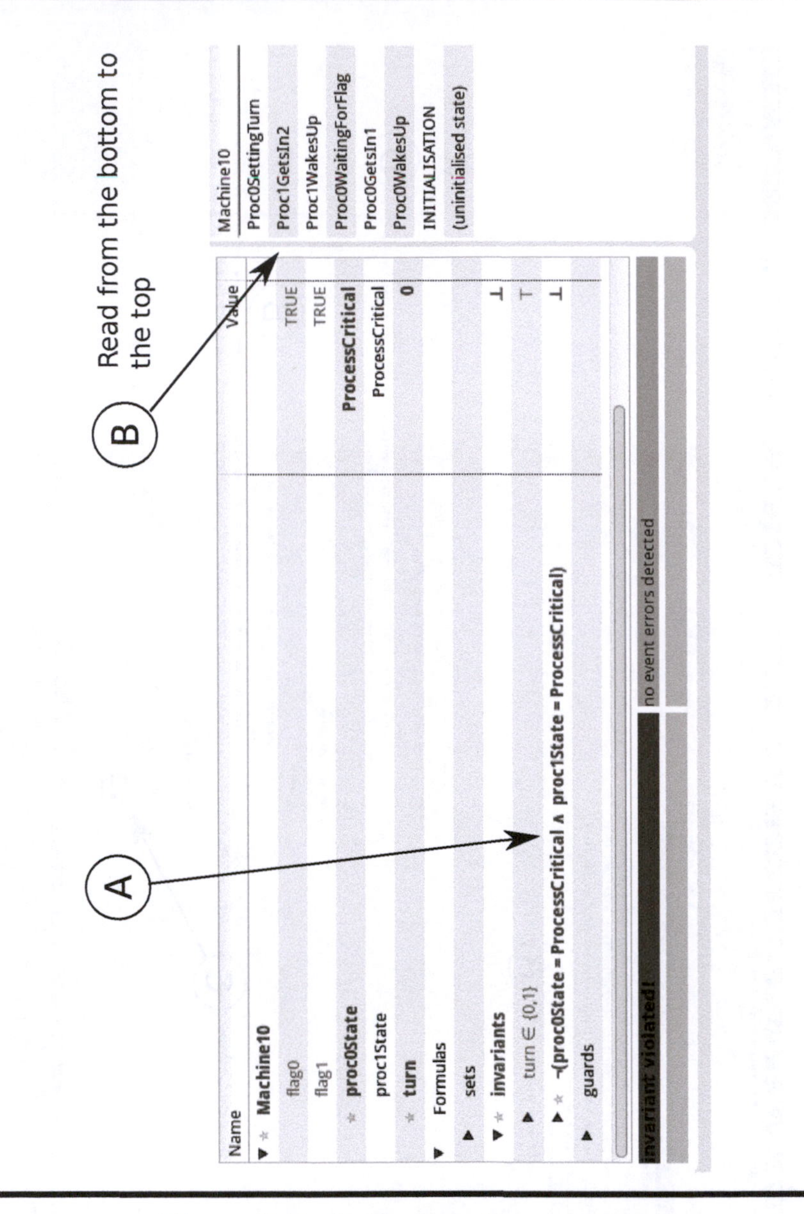

Figure 15.7 Rodin finds the violation.

Proof Obligations
INITIALISATION/inv3/INV
INITIALISATION/inv8/INV
INITIALISATION/act3/FIS
Proc0WakesUp/inv8/INV
Proc0GetsIn1/inv8/INV
Proc0Leaves/inv8/INV
Proc1WakesUp/inv8/INV
Proc1GetsIn1/inv8/INV
Proc1Leaves/inv8/INV
Proc1GetsIn2/inv8/INV
Proc0GetsIn2/inv8/INV
Proc0SettingTurn/inv3/INV
Proc0SettingTurn/inv8/INV
Proc1SettingTurn/inv3/INV
Proc1SettingTurn/inv8/INV
Proc0WaitingForFlag/inv8/INV
Proc1WaitingForFlag/inv8/INV

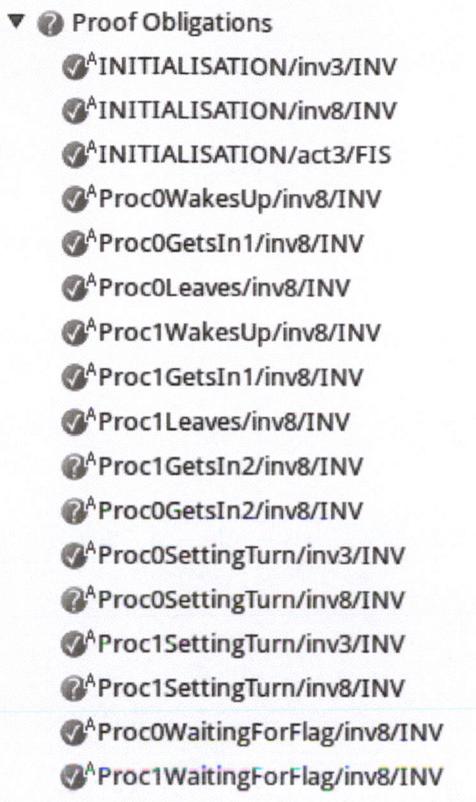

Figure 15.8 Rodin proves all but two invariants.

CODING

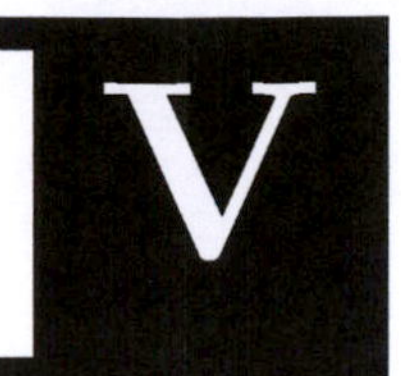

Chapter 16

Coding Guidelines

This part of the book deals with a few of the techniques "recommended" or "highly recommended" in the tables in ISO 26262 and IEC 61508 for the implementation rather than the design or verification of a product.

Table 1 of ISO 26262-6 and Tables A.3 and C.3 of IEC 61508-3 contain recommendations about programming languages and guidelines.

Programming Language Selection

It is unusual for designers and programmers to be given a free choice of programming language. Often there is a legacy system already coded, on which the new system is to be built; normally, the development company will have support for only a few languages, and some languages are simply not suitable for some tasks.

Frankly, there is no good programming language for writing the code for safety-critical systems, particularly for low-level, embedded systems. To some extent, this is a human problem: We are not wired for writing large numbers of unambiguous statements in poorly defined languages. Also we are victims of our drive for efficiency — for most (nonsafety) applications we would prefer our languages to be easily optimized for space or speed than for being easily demonstrated correct. A pointer in the C language, for example, is very efficient and very dangerous.

There seem to be two ways out of this *impasse*: to generate much more code automatically from a formal design (see Chapter 15), and to deploy much better static analysis tools (see page 265) to check the code still produced by humans.

In the meantime, the best we can do is choose a suitable language for the application being produced, define an acceptable subset of that language, and force that subset onto possibly unwilling programmers by rigorous code inspections, preferably carried out by an automated tool.

Programming Languages and the Standards

Table C.1 of IEC 61508-7 lists some programming languages and says whether they are recommended or not for use in safety-critical applications. ISO 26262 is less explicit, simply providing a list of characteristics that a programming language needs (5.4.6 of ISO 26262-6).

Language Features

The features of a programming language can help or hinder its use in developing a safe product. Commenting in 1981 (reference [1]) on the then-ongoing definition of the Ada language, C.A.R. Hoare said:

> *At first I was extremely hopeful. The original objectives of the language, included reliability, readability of programs, formality of language definition, and even simplicity. Gradually these objectives have been sacrificed in favor of power, supposedly achieved by a plethora of features and notational conventions, many of them unnecessary and some of them, like exception handling, even dangerous. We relive the history of the design of the motor car. Gadgets and glitter prevail over fundamental concerns of safety and economy.*

I list below a few of the characteristics that a programming language needs to be used to create safe systems and even, timidly, disagree with Hoare and the standards on the use of exceptions.

Full Definition

One major positive characteristic is an unambiguous and complete definition of the language. In this regard, a language such as C falls down badly, with a large number of syntactically correct statements being undefined in the C specification. For such expressions, a compiler is allowed to generate any code! Examples of such ambiguity include:

```
int x[4] = {0, 1, 2, 3};
int i = 2;

int *p = x + 5;        /* undefined */

x[i] = i++;            /* undefined */

i = 1 / x[0];          /* undefined */
```

It could also be argued that a language within which

```
if (x = 16) { .... }
```

is syntactically correct but does not mean "if x is equal to 16 ..." is perhaps unsuitable for dependable applications.

Explicit Dependability Support

Some languages, for example Eiffel, Ada 2012, RUST, and D, have explicit support for programming by contract (see page 24), and older languages, such as ALGOL, support recovery blocks (page 117). Both of these techniques can be useful for writing safe code. Although it is not difficult to implement programming by contract in other languages (packages exist for C++, Java, Python, and Ruby), it is more convenient to have this as a built-in language feature.

Predictable Timing

Real-time applications (see page 23 for the meaning of this term) may need to behave predictably so that timing guarantees can be met. Today, this strong guarantee is weakening in some designs, being replaced by a probability of the system not meeting its deadlines, but it is still required in many systems. In the past, languages that provided automatic garbage collection* could not provide this predictability because the timing of the garbage collection was not known in advance. Today garbage collection algorithms are more sophisticated and can be initiated at convenient times. The unpredictability of garbage collection

* "Garbage collection" is the automatic reclaiming of memory no longer required by applications, thereby removing from the application developer the overhead of having to remember to free it explicitly.

now has to be balanced against the significant reduction in memory leaks and other faults associated with manual memory management.

Anecdote 22 *I once worked on a system with real-time constraints written in two dialects of LISP. From time to time, while garbage collection took place, the application would be suspended for so long that, when it was awoken, it assumed that an operator had changed the system clock and went into its clock-resetting routine.*

Exception Handling

There is debate about whether an error return from a routine should be handled by an exception hierarchy or by passing return codes from the server back to a client. I am strongly of the opinion that, if the programming language supports exception handling, the disadvantages of return codes mean that exceptions are the only way to signal errors in a safety-critical application.

Return codes have to be values that could not be returned in the non error case, and this means that they differ from function to function. But the primary weakness of using return codes is that the server loses control of the exchange — it cannot guarantee that the client will check the return code.

However, section C.2.6.2 of IEC 61508-7 specifically "suggests" that coding standards be deployed to discourage the use of "Ada or C++-like exceptions." Although I am reluctant to argue with C.A.R. Hoare (see the quotation above) and somewhat less reluctant to argue with IEC 61508, I believe that discouraging the use of exceptions is wrong — a well-structured exception hierarchy is a more controlled environment than a free-for-all of return codes that may not be checked.

Suitability for Static Checking

With testing becoming less effective as the size of the state space of programs becomes astronomical, static checking is becoming more important. It would be nice if programming languages made static checking more efficient, but, for most languages, this was not a goal in their design.

At one end of the spectrum, C's extensive use of pointers, particularly function pointers, makes static analysis very difficult. At the other end, some languages are particularly amenable to such checking. The Erlang language, for example, only allows the value of a variable to be set once and it cannot thereafter be changed (so it's really not a variable). This makes static analysis much more efficient than with more conventional languages.

Significant Field Use

Demonstrating that the tool-chain used to preprocess, compile, assemble, and link a safety-critical device's source code is not easy — see page 293. One of the arguments that will probably form part of the safety case is that the level of field usage provides a certain level of confidence in a compiler. For the GNU C compiler, a vast history of usage and error reports exist; for other languages, such as D and RUST, that level of "confidence from use" may not yet be available.

Enforcement of Strong Typing

There are two concepts that are sometimes confused: strong typing and dynamic typing.

A language has strong typing if a variable cannot be used in a manner inappropriate for its type. Python, for example, provides strong typing and prevents the addition of an integer to a string:

```
>>> x = "5" + 6
Traceback (most recent call last):
  File "<stdin>", line 1, in <module>
TypeError: cannot concatenate 'str' and 'int' objects
```

The equivalent program in C compiles and executes without an error message because C is weakly typed:

```
int main()
    {
    printf("%s\n", "5"+6);
    return 0;
    }
```

The standards agree that weak typing is generally not good in a language being used for safety-critical applications.

The question of dynamic and static typing is more ambiguous.

Python, for example, allows the type of a variable to change dynamically:

```
>>> x = "hello"        # x has type string
>>> print x + " Chris"
hello Chris
>>> x = 8              # x has a numeric type
>>> print x + 32
40
```

whereas C does not allow the type of a variable to be changed without casting:

```
int main()
    {
    char  x[] = "hello";
    printf("%s\n", x);
    int    x = 8;
    printf("%d\n", x+32);
    return 0;
    }
>> gcc tryit.c
tryit.c: In function main:
tryit.c:7:11: error: conflicting types for x
      int   x = 8;
```

Dynamic typing allows code to be more compact and, in particular, to have fewer conditional statements. Conditional statements are the source of additional test cases, and, empirically, it has been found that conditional statements are the most likely types of statements to contain faults. It can be argued that avoiding them by using a language with dynamic typing is worthwhile.

The standards, however, look for both strong and static typing, so the use of a language with dynamic typing would have to be justified.

Use of Language Subsets

Both ISO 26262 and IEC 61508 strongly recommend the use of "language subsets," placing restrictions on the use of various language features so as to avoid some of the more misleading or ambiguous parts of the language and to reduce the possibility of runtime errors. These restrictions are normally defined in a coding standard.

Table C.1 of IEC 61508-7 lists some programming languages (although, to my disappointment, MODULA 2 and PL/M were removed in the 2010 version), and no complete language is highly recommended for SIL 3* or SIL 4 applications unless a subset is defined and used. Although not *strongly* recommended, it is interesting to note that Ada, Pascal, and FORTRAN 77 are recommended in their full (rather than subset) form for use in SIL 3 and SIL 4 applications. Because of its garbage collection, Java is not recommended even when constrained to a subset.

Neither IEC 61508 nor ISO 26262 provides examples of acceptable coding standards, because they are obviously language-specific and, to some extent, application-specific. ISO 26262 does, however, use the MISRA C** as an example of a coding standard for the C language, and section C.2.6.2 of IEC 61508-7 contains suggestions of the types of restrictions that should be considered when writing a coding standard.

Coding standards have never been popular with programmers, who often feel that they are being made to write code while wearing a strait-jacket. "It's like asking Picasso to paint a picture while not being allowed to use yellow paint." For example, in the C language, many errors occur when signed and unsigned integers are mixed within an expression, and it is not unreasonable to specify a language subset where such constructions are forbidden. This immediately leads to declarations such as

```
unsigned x = 0;
```

being flagged as coding standard violations because x is unsigned, whereas 0 is signed. It is this over-strict interpretation of the language subset that frustrates programmers.

An interesting shift can be detected between the 2004 and 2012 MISRA C standards — a move towards a greater proportion of rules that can be automatically checked ("decidable" rules), even if that has meant "softening" the rule slightly. I have heard programmers say, "I don't care whether there are 10 or 1000 coding rules, as long as they are all checked for me automatically."

In the past, there has been a tendency for coding rules to relate to a single source file; now, the MISRA C standard has introduced system-

* For a description of the different safety integrity levels (SILs), see page 36.

** See `http://www.misra.org.uk/`

wide coding rules. For example, "The same file shall not be open for read and write access at the same time on different streams."

So, What Is the Best Programming Language?

The question of what is the best programming language to use for the development of a safety-critical embedded device is a perennial topic in both Internet and face-to-face discussion groups. My personal choice would perhaps be D or RUST, particularly if confidence can be generated in the compilers.

References

1. C. A. R. Hoare, "The Emperor's Old Clothes," *Commun. ACM*, vol. 24, pp. 75–83, Feb. 1981.

Chapter 17

Code Coverage Metrics

This chapter describes some of the challenges and problems associated with gathering code coverage metrics. Loosely described, a coverage metric is a measure of how much of the program code has been exercised (covered) during testing.

In this chapter, I deal with coverage metrics from two points of view. The first section deals with the types of code coverage that provide some indication of the quality of the testing carried out: "Our testing has exercised every line of code and executed 95% of branch instruction in both directions."

The second section, starting on page 257, explains how the number of test cases needed to test combinations of parameters can be reduced without significantly affecting the quality of the tests; that is, how to achieve good combinatorial coverage.

Code Coverage Testing

One problem that plagues all forms of testing is deciding when to stop — when has the testing been sufficient? One way of assessing this is to set a target (say 100%) and then measure what percentage of the code has been executed at least once during testing. When the target is reached, testing can stop.

I deliberately used an ambiguous term in the previous paragraph: "code." Does this refer to the lines in the source code (e.g., C), to lines of the preprocessed source code, or to statements in the compiled object code? We will return to this important question later.

The idea of collecting any form of code coverage metrics is based on two assumptions:

1. It is potentially dangerous for lines of code to be executed for the first time in a safety-critical product in the field; all code should at least have been executed in the laboratory during testing.
2. The coverage metrics provide at least secondary evidence for the effectiveness and completeness of the testing.

The first of these is irrefutable; the second is more dubious.

Types of Code Coverage

Entry Point Coverage

This is the least strict form of coverage, measuring the proportion of functions that have been entered at least once during testing. It would be unreasonable to ship a software product to be used in a safety-critical application without achieving 100% entry point coverage.

Statement Coverage

Statement coverage measures the proportion of statements that have been executed at least once during testing. It is unlikely that 100% statement coverage will be achieved, primarily because of defensive coding. For example, consider the following code snippet:

```
typedef enum color { red, green, blue } color;
color x;
...
switch (x) {
    case red:
        ...
        break;
    case green:
        ...
        break;
    case blue:
        ...
        break;
    default:
        error("Bad value");
    }
```

A default clause has been included in the switch statement, even though the `case` clauses are exhaustive. This is good defensive coding, protecting against a programmer adding an extra color to the `enum` later and forgetting to update this `switch` statement. It is, indeed, required by the 2012 MISRA-C standard. However, the default clause can never be reached during testing.

Branch Coverage

Branch coverage is poorly defined, both in the standards and in the literature. For a simple branch statement of the form

```
if (x > 7) ....
```

100% branch coverage can be achieved by running tests where the condition was true at least once, and false at least once.

With a more complex condition, what is needed to achieve 100% coverage is less well defined. Consider the following conditional statement:

```
if ((x > 7) || (y < 23)) .....
```

Some references indicate that it is necessary only to make the entire, compound condition true at least once and false at least once. Thus, the two test cases, $x = 8, y = 45$ and $x = 3, y = 45$, are enough to achieve 100% branch coverage. This does not, however, demonstrate y's role in the condition, it being unchanged between the test cases.

The popular, open-source tool `gcov` measures coverage, including branch coverage, on C code compiled with particular options to force the compiler to insert the necessary statistics-gathering instructions. Its way of counting branches is more complex than the simple Boolean did-branch/didn't-branch metric described above.

MC/DC Coverage

MC/DC* stands for "modified condition/decision coverage," and this is the most stringent of the coverage conditions. The parts of the term can be built up as follows.

---

* Although this is normally read as "em-see, dee-see," it would be more accurate to read it as "em, see-dee, see."

Decision coverage.
This is a combination of entry point coverage and simple branch coverage.

Condition/decision coverage.
This is a combination of entry point coverage, simple branch coverage, and a requirement that all elementary Boolean conditions in every branch condition have been true and have been false at least once. In the above example of the compound `if` statement, it would no longer be adequate for y to remain unchanged. However, two test cases would still suffice: $x = 8, y = 10$ (both true) and $x = 3, y = 35$ (both false).

Modified condition/decision coverage.
This includes everything in condition/decision coverage, and adds the additional requirement that each elementary Boolean condition must be shown to affect the branching, all others being held constant. Now, two test cases no longer suffice to test the compound `if` statement above. Additionally we must demonstrate that, with x being held constant, the value of y can affect the result, and *vice versa*. For this example, 4 test cases will be required to satisfy the MC/DC conditions, but 2^N test cases are not always required, as is illustrated in the next example.

To compare the test cases required for branch and MC/DC coverage, consider the `if` statement:

```
if ( (x == 7) && ((y == 8) || (z == 23)) ) ...
```

Only 2 test cases are needed for simple branch coverage: one with $x = 4, y = 8, z = 23$ and one with $x = 7, y = 8, z = 23$. The values of y and z are largely irrelevant as long as at least one of them is true.

For MC/DC coverage, it might be thought that $2^3 = 8$ test cases are required, but this is not so. Consider the 4 test cases:

Test Case	x	y	z	x == 7	y == 8	z == 23	Outcome
1	7	8	25	True	True	False	True
2	8	8	25	False	True	False	False
3	7	9	25	True	False	False	False
4	7	9	23	True	False	True	True

The overall outcome of the compound conditional statement has been true at least once and false at least once, so simple branch coverage is 100%. Each of the three elementary Boolean expressions has been true and each has been false. Additionally, each has been varied while holding the other two constant and has been shown to affect the outcome of the compound condition.

I have presented a slightly simplified version of MC/DC testing here. Reference [1] describes three different types of MC/DC testing.

Basis Path Coverage

The flow of control through a program can be expressed in the form of a graph, where each node represents a decision point (e.g., `if` statement) and each arc represents a sequence of instructions without branching.

For example, the following function calculates the greatest common divisor (GCD) of two positive integers greater than zero by using Euclid's algorithm (I have added line numbers for convenience):

```
 3 int euclid(int m, int n) {
 4      int     r;
 5      if (n > m) {  // set m >= n
 6          r = n;
 7          n = m;
 8          m = r;
 9      }
10      r = m % n;
11      while (r != 0) {
12          m = n;
13          n = r;
14          r = m % n;
15      }
16      return n;
17 }
```

This function can be expressed as a graph, as shown in Figure 17.1. The numbers in the nodes of the graph represent the line numbers in the program, and the numbers on the arcs are for convenience in the description below. It can be seen from the graph that the first decision point is whether to go to line number 6 or 10 (depending on the `if` statement on line 5).

It is from such a graph that the McCabe cyclomatic complexity of a program is calculated. The measure is named after Thomas McCabe, who first proposed it in reference [2]. The McCabe complexity is calculated by

$$M = E - N + 2P \tag{17.1}$$

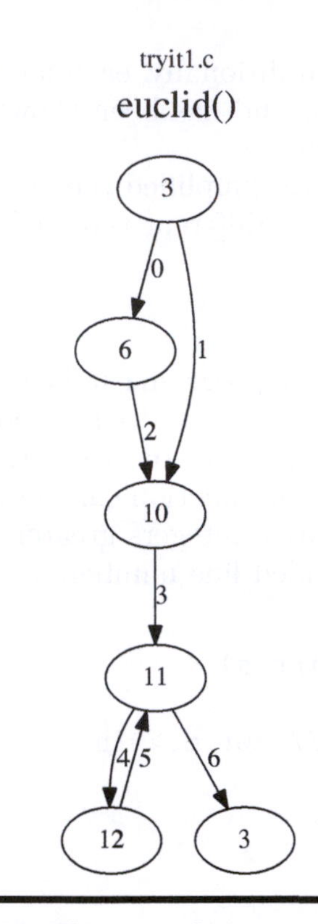

Figure 17.1 Flow graph for sample program.

where E is the number of edges (arcs) in the graph ($E = 7$ in Figure 17.1), N is the number of nodes ($N = 6$)* and P is the number of disconnected parts of the graph ($P = 1$). Thus, for the program in Figure 17.1, the cyclomatic complexity = 7 - 6 + 2 = 3. Coding and design standards often call for all functions to have complexities no greater than 10, but this may vary from project to project. As reference [3] states:

> *The original limit of 10 as proposed by McCabe has significant supporting evidence, but limits as high as 15 have been used successfully as well. Limits over 10 should be reserved for projects that have several operational advantages over typical projects, for example experienced staff,*

* Note that the node for line number 3 is repeated to "close" the graph.

> *formal design, a modern programming language, structured programming, code walkthroughs, and a comprehensive test plan. In other words, an organization can pick a complexity limit greater than 10, but only if it is sure it knows what it is doing and is willing to devote the additional testing effort required by more complex modules.*

In addition to being the input for the estimation of complexity, the flow graph also provides the definition of "basis" path coverage. The McCabe cyclomatic complexity of the program is not only $E - N + 2P$, it is also the number of test cases that will be required to perform 100% basis path coverage.

A "basis" for a vector space is a set of vectors, none of which can be obtained by taking linear combinations of the others. Most of the tools that generate basis paths hide the mathematics from the user, but if you are interested, see page 333. For generating test cases, the vectors are the potential paths through the program:

Path	Vector: Arcs Covered							Arcs
	0	1	2	3	4	5	6	
Path A	0	1	0	1	0	0	1	1 → 3 → 6
Path B	1	0	1	1	0	0	1	0 → 2 → 3 → 6
Path C	0	1	0	1	1	1	1	1 → 3 → 4 → 5 → 6
Path D	1	0	1	1	2	2	1	0 → 2 → 3 → 4 → 5 → 4 → 5 → 6

So, path A (1 → 3 → 6) can be represented by the vector (0, 1, 0, 1, 0, 0, 1) and path B by the vector (1, 0, 1, 1, 0, 0, 1).

Because of the loop between lines 12 to 14, there is potentially a very large number of paths through this program (limited only by the size of an integer on the computer on which it is executed), and path D is a path where the loop has been traversed twice.

However, it is easy to check that

$$D_i = B_i - 2 \times A_i + 2 \times C_i \quad \text{for} \quad 0 \leq i < 7 \tag{17.2}$$

for each of the 7 elements of paths A to D. Thus, Path D is not linearly independent of paths A to C.* In fact, any path through the graph in Figure 17.1 can be expressed as a linear combination of paths A, B and C. Thus, the three paths A to C (or, if you prefer, paths A, B and D)

* This is one of McCabe's original examples — see reference [3].

form a *basis* in the 7-dimensional path space, and three tests can be chosen to exercise those paths to achieve 100% basis path testing.

Full basis path coverage implies full branch coverage.

Path Coverage

For anything but the simplest program, full path coverage cannot be achieved. Given a control flow graph of the type shown in Figure 17.1, the number of potential paths is limited only by Euclid's algorithm and the size of an integer on the computer executing the program. For example, the loop between lines 11 and 12 is repeated 25 times if the coprime values $m = 317811, n = 196418$ are used.

Coverage and the Standards

The second edition of IEC 61508-3 (Table B.2) calls for targets of 100% entry point coverage, branch coverage, and MC/DC coverage. It acknowledges that defensive code of the type illustrated in the switch statement in the code on page 248 cannot be reached, but requires that, for coverage level below 100%, an "appropriate explanation should be given."

While not setting target levels, Table 12 of ISO 26262-6 also recommends the collection of statement, branch and MC/DC coverage metrics.

The difference between the two standards lies in the phase when each expects the coverage to be measured. ISO 26262-6 is clear: Table 12 relates to the software unit level, and it is in the section relating to software unit testing. IEC 61508 is more flexible, indicating that the coverage statistics are to be collected during either module or integration testing.*

While achieving a high level of coverage during white-box** module testing is easier than during black-box integration testing, a shortfall in coverage during integration testing may, in some instances, indicate incompleteness in the integration test cases. If, for example, a particular branch, taken only if memory exhaustion occurs, is not taken during integration test, then this might be an indication that additional integration testing should be added to cover the low-memory condition.

* Note that various standards use the terms "module test" (IEC/ISO 29119 and IEC 61508) and "unit test" (ISO 26262 and IEC 62304) synonymously.

** "White-box" testing is performed with a knowledge of how the module is implemented; "black-box" testing is performed with knowledge only of the function's external interface.

IEC 62304 also makes passing reference to code coverage in Appendix B.5.7, where it indicates that it is an activity to be carried out during the system testing phase.

Effectiveness of Coverage Testing

When we perform a test, we actually need to achieve *fault* coverage — that is, we want to execute tests that expose *faults*. However, the number of faults in the code is unknown, and so measuring the level of fault coverage is impossible. Instead, the level of branch or MC/DC coverage is used as a proxy for fault coverage, and a reasonable question is, "Does x% of branch or y% of MC/DC coverage correlate with a particular level of fault coverage?"

This question is addressed in reference [4] by Loïc Briand and Dietmar Pfahl and reference [5] by Yi Wei *et al.* The outcomes of those studies are depressing:

> *The study outcomes do not support the assumption of a causal dependency between test-coverage and defect-coverage.*

and

> *These results ... also show that branch coverage is not a good indicator for the effectiveness of a test suite.*

from the two papers, respectively. In part, this is to be expected. Consider the simple program

```
int v[100];
void f(int x) {
    if (x > 99) x = 99;
    v[x] = 0;
    return;
    }
```

where the programmer has tried (but failed!) to guard against an out-of-range input value. Two test cases suffice to provide 100% statement, branch, MC/DC, and basis path coverage: $x = 50$ and $x = 101$. Unfortunately, neither of these catches the fault that is revealed when

$x = -4$.

When I asked the KLEE tool (see page 273) to generate test cases for that snippet of code, it returned three test cases: x = -7387680 (which exposes the bug), x = 2147483644 (which is larger than 99!) and, x = 97. Only two of these are strictly necessary for complete code coverage.

I hinted at another problem associated with coverage metrics earlier: What do we count as source code lines? Do we count the source code lines as written by the programmer, the source code lines after preprocessing (either by the standard preprocessor or by a specialist pre-processor, such as that injecting coded processing commands — see page 138), or machine code "lines" following compilation?

Optimizing compilers can remove, re-order and insert lines between their source input and their assembler output. These transformations make statement and branch coverage hard to measure: perhaps a set of tests executes 100% of the branches in a C source program, but only 40% of the branches in the generated assembler.

The C language concepts of `inline` code and macros can also lead to unreasonable coverage values. If a piece of code that is declared to be `inline` contains a branch, then that branch may be taken in some instances where the code is inserted, but not in others. If the analysis tool measuring coverage sees the code after the inlining has occurred, it will report uncovered code, although none exists. The same is true for macros.

Achieving Coverage

Various tools exist to generate test cases that can at least provide good statement and branch coverage. One such tool is the open-source KLEE symbolic execution program described on page 272. As it carries out its symbolic execution, it performs the analysis necessary to determine the values of the variables needed to drive the program down each of its paths. When KLEE is applied to the program calculating the greatest common divisor (page 251), it generates the following test cases:

- $m = 2147483647, n = 2147483647$ (GCD = 2147483647).
- $m = 50790404, n = 54463393$ (GCD = 1).
- $m = 679738981, n = 340287488$ (GCD = 1).

Notice that the first test case corresponds to Path A in the table on page 253, the second to Path D (with 18 rather than 2 iterations of the loop around the lines 11 and 12, and the third to Path C (with 17 times around the loop). These three test cases form a basis and so provide 100% entry point, statement, branch, and basis path coverage. Because of the simple nature of the decisions, they also provide 100% MC/DC coverage. On my laptop computer, KLEE took a few seconds to deduce them.

It should be said that, while automatically generated test cases can provide excellent coverage of various types, in my experience these tests rarely find genuine bugs.

Combinatorial Testing

This section deals with coverage in a slightly different manner by asking the question, "If a function or utility program takes several parameters, then what level of coverage is required to test it adequately?"

Consider our fictitious company Beta Component Incorporated (BCI), as described in Chapter 4. This company produces an operating system that Alpha Device Corporation (ADC) intends to use in its device. An operating system is a program that can potentially take a number of different command-line parameters. Examples might include:

- Set timer interrupt period (4 values: $500\mu s$, 1ms, 2ms, 10ms)
- Enable kernel pre-emption (Yes/No)
- Select required file system (4 options)
- Detect memory alignment faults (Yes/No)
- Use software floating point emulation (Yes/No)
- Disable the processor actually halting when in the idle thread to allow faster exit from the idle state (Yes/No)
- Defragment memory automatically (Yes/No)
- Don't lock, lock, or superlock memory pages (3 values)
- Enable address-space randomization to increase security (Yes/No)
- Set the debug verbosity level (5 values)
- Enable backward compatibility with BCI's previous operating system (Yes/No)

In total, there are $4 \times 2 \times 4 \times 2 \times 2 \times 2 \times 3 \times 2 \times 5 \times 2 = 30720$ different ways in which this component can be configured. In practice, this computation might be more complex because certain combinations might not be permitted; for example, verbosity level 3 might only be applicable if floating point emulation is also selected.

To test this component thoroughly, BCI needs to run a set of regression tests on all 30,720 combinations of input parameters. A regression test of an operating system typically takes hours or even days. If BCI's regression suite takes 24 hours to execute, this level of testing would take a total of about 84 years to complete. Of course, in practice, several tests would be executed in parallel on different hardware, but even if 20 such systems were available, the test duration would still be more than 4 years.

Combinatorial testing, as described in numerous papers and the book [6] by D. Richard Kuhn, Raghu Kacker, and Yu Lei, is based on a significant observation: When a program has many parameters, then exhaustive combinations of a few of them tend to uncover almost all problems. This is an empirical rather than theoretical observation and is backed in reference [6] by the comparison of bugs uncovered by exhaustive and combinatorial testing of a number of different programs.

BCI's operating system as listed above has 11 parameters, and so an exhaustive test requires all possible combinations of all possible values of all 11 parameters to be tested. And we know that this will need 30,720 tests.

The question can be posed as to how many test cases will be required to test all possible combinations of all possible values of $t \leq 11$ parameters. Unfortunately, there is no known answer to this question, but the number of test cases is known to be of order

$$\nu^t \log(N) \tag{17.3}$$

where ν is the number of possible variables a parameter can take (in our example that's bounded by the debug verbosity to 5), t is the "strength" of the testing (all combinations of t of the parameters), and N is the total number of parameters (11 in the example).

Although there is no closed-form solution for how many test cases are required, there are resources to help select test cases and determine how many test cases will be required. In particular,
`http://www.public.asu.edu/~ccolbou/src/tabby/catable.html`
provides a list of the best values for the number of test cases known to date. In our example, even if we assume that every parameter can take 5 values, we arrive at the maximum number of test cases listed in Table 17.1.

Table 17.1 can be read as follows: If each of the 11 parameters that configure the operating system had 5 possible values (for a total of 48,828,125 possible tests, were the system to be tested exhaustively), then fewer than 1245 test cases are required to test all possible combinations of every 4 of those parameters.

Table 17.1 Bounds on numbers of test cases ($\nu = 5, N = 11$).

Strength t	**Bound on number of test cases**
2	37
3	225
4	1245
5	8,554

These are still large numbers of tests, but they are known to be very much larger than will actually be required, because most of the parameters are Boolean and therefore only take 2 values. However, if we have 20 test systems in the laboratory, even having 1245 test cases reduces the test time from 4 years to 2 months — it's moving into the realm of possibility.

There are several tools that can be used to find a better estimate of the number of tests required and a listing of the parameter values to be used for those tests. One such is ACTS, which can be downloaded from the NIST website at
`http://csrc.nist.gov/groups/SNS/acts/index.html`.

ACTS does not claim to find the lowest possible number of test cases (this number is not known), but in practice, it finds reasonable test cases that at least provide the specified coverage.

Table 17.2 Actual number of test cases.

Strength (t)	**Number of test cases**
2	20
3	82
4	260
5	693
6	1923

Table 17.2 lists the minimum number of test cases found by ACTS for the real values of BCI's command-line parameters. The 1923 test cases needed to include all combinations of all values of every 6 of the 11 parameters are too numerous to include here (the ACTS tool is free — download it and get the results yourself!), but for convenience, I have included the 20 $t = 2$ test cases in Table 17.3; I have abbreviated the column headings to get the table to fit on the page, but they align with the list on page 257.

Table 17.3 The strength $t = 2$ test cases.

Tmr	P/E	FS	Aln	SWfpt	Hlt	Defrg	Lck	Rnd	Vrb	B
500	F	2	F	F	F	F	2	F	1	F
1ms	T	3	T	T	T	T	3	T	1	T
2ms	F	4	T	F	T	F	1	T	1	F
10ms	T	1	F	T	F	T	2	F	1	T
500	T	3	F	F	T	T	1	F	2	F
1ms	F	4	T	T	F	F	2	T	2	T
2ms	F	1	F	F	F	T	3	F	2	F
10ms	T	2	T	T	T	F	1	T	2	T
500	T	4	F	T	F	T	3	T	3	F
1ms	F	1	T	F	T	F	1	F	3	T
2ms	T	2	F	T	T	T	2	F	3	T
10ms	F	3	T	F	F	F	3	F	3	F
500	T	1	T	T	F	T	1	T	4	T
1ms	F	2	F	F	T	F	3	F	4	F
2ms	F	3	T	F	T	F	2	F	4	T
10ms	T	4	F	F	F	F	3	F	4	T
500	T	1	T	T	F	T	3	T	5	T
1ms	F	2	F	F	T	F	1	F	5	F
2ms	F	3	T	T	T	T	2	T	5	T
10ms	T	4	T	F	T	F	2	F	5	T

With $t = 2$ (i.e., all possible combinations of every *pair* of parameters), there must be at least $5 \times 4 = 20$ test cases and, in Table 17.3, ACTS has found a set of precisely 20 test cases. For $t = 3$, there must be at least $5 \times 4 \times 4 = 80$ test cases. Table 17.2 indicates that ACTS failed to find such a set — but it came close with a set of 82 test cases.

In practice, it is rarely necessary to set $t > 6$, and with $t = 6$, the 1923 test cases, using the test durations above, should take about 3 months to execute. This is significantly better than the 4 years calculated previously.

The testing of large numbers of combinations is not only a problem for BCI. Alpha Device Corporation manufactures devices, incorporating BCI's operating system, that are installed in cars. Assume that these devices may be fitted in any of 8 types of car, may have a large or small memory option, may be ruggedized or not, may work with any of 6 types of display, and may run on any of 4 processor chips. That means that ADC has $8 \times 2 \times 2 \times 6 \times 4 = 768$ different versions of each device to test. Again, combinatorial testing may be a useful tool: With

$t = 2$, ACTS finds a set of 48 test cases and, with $t = 3$, it finds a set of 192 test cases. The gain is not as great as with BCI's example, but it is still possibly worth having.

The message of combinatorial testing is clear: If the number of combinations of parameters makes a full test suite impractical, then the subset of test cases that it is useful to run can be chosen scientifically to provide good coverage. This subset is almost certainly better than any subset chosen, effectively at random, by a tester.

Summary

It is difficult to argue that it is acceptable that a section of a program in a safety-critical device is executed for the first time in the uncontrolled environment of the real world, rather than in the controlled environment of a test laboratory. For this reason, some level of coverage needs to be achieved during testing.

However, code coverage metrics are often seen as an easily measurable proxy for fault coverage — useful as an end in themselves. That claim I do not find convincing.

References

1. Office of Aviation Research Washington, D.C., USA, "An Investigation of Three Forms of the Modified Condition Decision Coverage (MCDC) Criterion," 2001. Available from `http://www.tc.faa.gov/its/worldpac/techrpt/ar01-18.pdf`.
2. T. J. McCabe, "A Complexity Measure," *IEEE Trans. Software Eng.*, vol. 2, no. 4, pp. 308–320, 1976.
3. A. H. Watson and T. J. McCabe, "Structured Testing: A Testing Methodology Using the Cyclomatic Complexity Metric," 1996.
4. L. Briand and D. Pfahl, "Using simulation for assessing the real impact of test coverage on defect coverage," in *Software Reliability Engineering, 2000. Proceedings. 10th International Symposium on*, pp. 148–157, IEEE, 1999.
5. Y. Wei, B. Meyer, and M. Oriol, "Is Branch Coverage a Good Measure of Testing Effectiveness?," in *LASER Summer School* (B. Meyer and M. Nordio, eds.), vol. 7007 of *Lecture Notes in Computer Science*, pp. 194–212, Springer, 2010.
6. R. Kuhn, R. N. Kacker, and Y. Lei, *Introduction to Combinatorial Testing*. London: CRC Press, 2013.

Chapter 18

Static Analysis

It is convenient to divide the verification of code into dynamic and static analysis. Dynamic analysis, simply called "testing" as a rule, means executing the code while looking for *errors* and *failures*. Static analysis means inspecting the code to look for *faults*.* Static analysis is effectively automated code inspection, where a program rather than a human performs the inspection.

As I have said in several places in this book, dynamic testing is becoming increasingly ineffective as the complexity of the tested system increases. Luckily, static analysis and formal verification (see Chapter 15) are increasing in scope to cover some of the shortfall left by dynamic testing. In part, this is due to increased processing power, but mainly it is due to the development of better algorithms.

The symbolic execution of code (see page 272) lies between dynamic and static analysis because it simulates the execution and, because it can handle many input values simultaneously, is more powerful than dynamic testing in some circumstances.

What Static Analysis Is Asked to Do

Static analysis of code has to balance several demands (this list is based on the one in reference [1] by Bruno Blanchet *et al.*):

- It must be sound in that it finds *all* problems within the code.

---

* The distinction between faults, errors, and failures is given on page 16.

This is what distinguishes static analysis from dynamic analysis — dynamic analysis can never be exhaustive.

- It must be automated, requiring little or no manual intervention to perform its analysis.
- It must be precise, avoiding false positives (unnecessary warnings).
- It must be scalable, being able to handle very large code bases.
- It must be efficient, limiting its use of memory and time.

This is a difficult set of requirements to meet, and most static analysis tools fail on the requirements of precision (they produce false positives) and efficiency (they take too long to run). Both of these weaknesses make them to some extent unsuitable for regular use.

A programmer editing and compiling a program needs to have the feedback from the static analysis tool in the same timescale as the compilation. If the static analysis adds more than a few seconds to the edit/compile cycle, then it will not be performed by the originating programmer. Instead the organization will perform it as a batch activity, perhaps running overnight. When this occurs, the direct link between the programmer writing the questionable code and the tool reporting it is lost, and examining the output from the static analysis becomes a chore.

Reference [2] by Benjamin Livshits *et al.* highlights the problem of balancing soundness (i.e., finding all problems in the source code) with the approximation that the tools need to take of the language:

> *Analyses are often expected to be sound in that their result models all possible executions of the program under analysis. Soundness implies the analysis computes an over-approximation in order to stay tractable; the analysis result will also model behaviors that do not actually occur in any program execution. The precision of an analysis is the degree to which it avoids such spurious results. Users expect analyses to be sound as a matter of course, and desire analyses to be as precise as possible, while being able to scale to large programs.*

In order to provide soundness, most if not all static analysis tools have to work with an approximation of the source language — perhaps not handling function pointers in C or reflexion in Java.

Static Code Analysis and the Standards

ISO 26262 recommends the use of static analysis as a mechanism for demonstrating the freedom from interference of two software elements (annex D of part 3). It also recommends it for verifying software implementation in Table 9 of part 3.

IEC 61508 recommends static analysis for software verification in Table A.9 of part 3 and gives information about the technique in paragraph B.6.4 of part 7, where it rather quaintly describes its purpose as "To avoid systematic faults that can lead to breakdowns in the system under test, either early or after many years of operation." It could be argued that the specific exclusion of a system that is past its early phase but has not been in the field for many years is unfortunate.

Static Code Analysis

It is tempting to think that, because programs are written in a formal language, it ought to be possible to write a program that will analyze a target program statically and answer the general question, "Does this program contain runtime errors?"

This is unfortunately impossible. In the first case, it is impossible to specify what a program *ought* to do. Consider this snippet of C code and ask yourself whether it contains a fault:

```
int  y = 10;
char x[8];
int  z = 7;

strcpy(x, "12345678");
```

Almost all programmers, and all static analysis tools, will immediately claim that there is a fault in the code: a beginner's fault. The program copies 9 bytes (the 8-byte string "12345678" and its terminating zero) to a field only 8 bytes wide. Depending on how the compiler has laid out the stack, either `y` or `z` will be corrupted.

This is true, but the purpose of the program was to allow the programmer to test an application for detecting and handling stack corruption — the program is doing exactly what it is intended to do in deliberately corrupting the stack. Without a formal definition of what the program is intended to do, it is impossible for either a human or a program to determine whether it is correct or not, and, generally, the only formal definition is the program itself.

However, there is a more fundamental problem. Alonzo Church, Kurt Gödel and Alan Turing proved that no deterministic mechanism can be executed by a computer to answer the question of whether all other programs are correct.

All static analysis is therefore a compromise. It finds many "false positives," reporting faults that do not actually exist, and often fails to detect faults that are present. The question is not whether static analysis of code can always be correct, but whether it can be useful.

Various levels of static analysis can be distinguished as in the sections below.

Syntax-Level Checking

The simplest form of static analysis is line-by-line syntax checking against the language specification and against local coding rules.

This type of checking is carried out by a compiler or by the class of tools known as "lint." Lint-like tools are available for most languages — lint for C and C++, pylint for Python, jsl for Javascript, lint4j for Java — and can be customised to check for the local coding rules and conventions.

The main disadvantage normally stated for this class of tool is the high level of "false positives" and pedantic "true positives." This is not unexpected, because line-by-line fault prediction is a hard problem to solve. The C statement on page 245:

```
unsigned x = 0;
```

to which very few programmers would object, does violate one of the MISRA C coding standards because it mixes numerical types (x is unsigned, 0 is signed). This statement would therefore be flagged as a problem by most static analysis tools, but I would classify it as a pedantic true positive.

To avoid false and pedantic positives, once each reported fault has been checked and found not to be genuine, some form of "grandfathering" has to be performed to prevent the fault being reported again.

In reference [3], Sunghun Kim and Michael Ernst analyze the output of three lint-like tools on historical versions of three programs to see whether the issues they detect represent actual faults. In summary, they found that only about 9% of the warnings given by the lint-like tools pointed to code that was changed as a result of bug fixing during the first one to four years of operation. This implies that 91% of the warnings related to code that, since it wasn't corrected, presumably didn't cause failures in the field. The analysis tools also prioritized their warnings, and it was found that only between 3% and 12% of the high priority warnings had been fixed. It is interesting to note that the authors of reference [3], having noted this lack of correlation between faults and fixes, go on to use a system based on the historical information in the bug database and code repository to improve the hit rate of the static analysis tool — mining information in a similar way to that proposed in Chapter 13.

However, in practice, the main disadvantage of syntax-level static analysis may be that given in the quotation about Picasso on page 245: While syntax errors obviously have to be fixed, reports about stylistic language use are perceived as being insulting: "It is useful for other programmers to be warned about unsafe language use, but clearly I don't need such warnings because I am a **real** programmer."

Semantic Enhancement to Syntax Checking

"Lint"-like syntax checkers as described above typically have very little semantic knowledge of the language being parsed. However, deeper problems can be found if the tool can be programmed in terms of language concepts such as "type," "path," "expression," "identifier," and "statement."

A C program could then be scanned for patterns of the type "a pointer expression is checked to be NULL and then dereferenced before being re-assigned" or "a variable declared to be unsigned is compared to see whether it is less than zero." This type of checking is either very difficult or impossible to perform in a purely syntactically way.

Tools such as Sparse and Coccinelle can handle this type of semantic checking, and can be particularly effective in uncovering potential security vulnerabilities in code (buffer overflows, etc.).

Sparse was specifically written to detect problems with the code in the Linux kernel, but I have used it with success on other operating system and library code.

Coccinelle* is a very useful tool, because it not only detects transgressions in the code, but can be scripted to produce *diff* files to correct the problems. For example, given the program in Figure 18.1, Coccinelle will, if provided with the appropriate script, create the following *diff* to correct the problem:

```
-    if (string[i] == NULL)
+    if (string[i] == '\0')
```

* http://coccinelle.lip6.fr/.

```
int doit(const char *string) {
    int i,j;

    for (i=0; i<strlen(string); i++) {
        if (string[i] == NULL)
            return 0;
        else
            j += 1;
        }
    return j;
    }
```

Figure 18.1 A program with an error.

The rub lies in the phrase "the appropriate script." The language in which Coccinelle scripts are written, SmPL, is precise and well-defined, but can be extremely awkward to write. However, for simple constructions it is straightforward: The following example is taken from the Coccinelle website** and detects the nonsense construction `!x&y` in the C language, where `y` is a constant. The Coccinelle script that detects this, but permits `!x & !y`, which may be meaningful, is readable even without a knowledge of SmPL:

```
@@ expression E; constant C; @@
(
  !E & !C
|
- !E & C
+ !(E & C)
)
```

As Coccinelle understands the concepts of an expression and a constant, it is very simple to define a script to remove `!E & C` and replace it with `!(E & C)`, which was almost certainly what the programmer meant.

The other use of tools such as Coccinelle is to prove a negative. Table B.1 in Annex B of IEC 61508-3, for example, calls for limited use of recursion and Table 8 of ISO 26262-6 bans recursion completely. How can we be sure that we have not used recursion somewhere in our codebase? How can we demonstrate this to an auditor?

** Copyright held by Gilles Muller and Julia Lawall.

It is difficult to prove a negative. One technique is to create a script for Coccinelle or an equivalent tool, demonstrate that it finds recursions, and then let it scan the entire codebase of the product without finding any recursive calls. A suitable script, prepared by Patrick Lee, is as follows:

```
// Detect a recursive call in C code

@rule1@
type t;
identifier id;
statement s;
function f;
position p;
@@
t id (...) {
...
id@p( ...)
...}

@script:python@
p << rule1.p;
x << rule1.id;
@@
print "%s Recursion %s on line %s" % (p[0].file,x,p[0].line)
```

Anecdote 23 *My colleague Patrick Lee once used this Coccinelle script to demonstrate that a large codebase did not include recursion. It demonstrated that the only recursion in the product was tail-recursion. He then justified to his auditor the use of tail-recursion on the grounds that it met the requirements of paragraph C.2.6.7 of IEC 61508-7: "If recursion is used, there must be a clear criterion which makes predictable the depth of recursion." This is obviously true for tail-recursion.*

In addition to open-source tools such as Sparse and Coccinelle, there are many commercial products that provide the same type of analysis, including XTRAN and Astrée. The theory underlying Astrée is given in reference [1].

Fault Density Assessment

Another useful outcome from static analysis, not only of the code itself, but also of the code repository, is identifying the modules and functions where the most faults are likely to lie. This allows those areas to be subjected to deeper scrutiny and verification.

This is similar to the mining of the code repository and problem database to estimate failure rates of new software, as described in *Assessing Failure Rates* on page 182, except that the aim is different — it is to analyze the existing code before shipment rather than predict the failure rate of code yet to be produced. For this reason, the number of factors taken into account can be much larger.

In general, tools searching for areas where faults are likely to be found take into account factors such as the programming language used, the level of control nesting, the variable density index (a ratio of the variable usage to the number of program statements), the density of comments in the code, the size of modules, the cyclomatic complexity modules, the number of global variables, the number of times each module has been changed, whether some or all of the code is imported from a third party (including open source), the age of the module, and the number of different programmers who have worked on each module.

An example of these (and other) factors being used to predict the risk associated with releasing a new version of a software system is provided in reference [4] by Audris Mockus and David Weiss.

Anecdote 24 *I once worked with an analyst who had studied the problem of identifying those modules likely to contain the most faults. He was always ready to bet a month's salary that he could identify the modules in any newly developed system that would require the most changes over the 12 months following product release.*

His model took 40 characteristics of the software system into account, and he made no attempt to understand why correlations existed between those characteristics and the fault density — in fact some of the correlations were strongly counter-intuitive.

To my knowledge, no one was ever brave enough to accept his bet.

There is some indication that the estimators used to predict the most fault-prone modules may vary between code developed using an agile and a waterfall process — see, for example, reference [5] by Thomas Ostrand and Elaine Weyuker. Certainly, within waterfall development processes, these techniques have been able to predict the 20% of modules that contain more than 80% of the faults, and this helps to focus inspection and testing resources. This is not to say that the technique cannot be applied to agile developments, just that different criteria probably dominate there.

Correctness Proofs Against Invariants

If a *programming by contract* approach is taken to code development (see description on page 24), then this can assist static analysis.

A routine to calculate the square root of a floating point number may include a precondition that the input value not be less than zero and a postcondition that the number returned, when squared, lies within a certain range of the input value. Internal invariants may also be specified. In some languages, for example Eiffel and D, these conditions are built into the language syntax; in other languages they may have to appear as structured comments. Where conditions cannot be included, the `assert()` statement can be used as a limited alternative.

Although these conditions are created to be checked at runtime, reference [6] by Rafael Ceballos *et al.* points out that they can be exploited by static analysis tools: "Is it possible in a given body of code that the square root routine could be presented with a negative number?"

Static analysis making use of constraints included for runtime analysis is not totally one-way traffic; static analysis can also check whether the constraints themselves are consistent and thereby provide value for the runtime checking.

This technique is described in reference [7] by David Crocker which demonstrates how preconditions and post-conditions can be added to a C++ program by means of macro definitions. However, Crocker points out that additional work is required to cover concurrency and floating point arithmetic. In the latter case, the tool has to recognize that, even if $x \neq 0$, it cannot be assumed that $x \times \frac{1.0}{x}$ is equal to 1.0.

One major drawback for checking against contracts, either at runtime or statically, is getting the contracts written into the code. This is very difficult to do retrospectively and really can only be done by the original programmer while the routine is being written. However, many programmers are allergic to adding this type of, in their mind, unnecessary code. To work around this reluctance, various automated "programming assistants," such as Daikon (reference [8]), have been proposed. Reference [9] by Michael Ernst *et al.* describes a technique of dynamically executing the code with random inputs and inspecting what values and relationships between values are common to all executions — those are likely to be the invariants.

Symbolic Execution

Advantages of Symbolic Execution

Symbolic analysis sits on the boundary between static and dynamic analysis. It is not purely static analysis because the code is actually executed: computations are performed, files are opened and read, etc. On the other hand, it is not purely dynamic analysis (testing) because the code is not fully compiled and is not executed as a stand-alone module with one set of input parameters.

One of the disadvantages of dynamic analysis (testing) is that each set of input variables requires a separate test case. As described in the section on combinatorial testing on page 257, this can often lead to very large numbers of test cases. In the case of an input variable that is defined to be a 32-bit integer, there are $2^{32} = 4,294,967,296$ possible values. The advantage of symbolic execution is that even such large ranges can be handled.

The symbolic execution engine can be thought of as a form of interpreter that reads and "executes" the code while keeping track of all the possible values that each variable can take at each point. Consider the following snippet of code:

```
int doit(unsigned x) // x is in the range 0 to 4294967295
    {
    unsigned   y;    // y is in the range 0 to 4294967295

    if (x > 256) {
        y = x / 2;   // x is in the range 257 to 4294967295
                     // y is in the range 128 to 2147483647
        }
    else {
        y = x / 3;   // x is in the range 0 to 256
                     // y is in the range 0 to 85
        }
    return y;
    }
```

When analyzing this, a dynamic test requires that a specific value be assigned to `x`, while a static analysis will not be able to check beyond the correctness of the variable types. A symbolic execution of the same code can preserve all the possible values that the variables could take at each instruction — as specified in the comments.

Symbolic Execution and the Standards

Table B.8 of IEC 61508-3 contains an explicit reference to symbolic execution, although the reference it makes to paragraph C.5.11 of part 7 indicates that the meaning of the term as used in IEC 61508 is somewhat less sophisticated than the tool described below.

KLEE Symbolic Execution Tool

Reference [10] by Cristian Cadar and Koushik Sen provides a survey of the application of symbolic execution up to 2013, and Cadar is the lead developer and maintainer of the KLEE tool* for symbolic execution.

This tool symbolically executes a semicompiled form of a program in the `llvm`** intermediate code, produces test cases to provide reasonable code coverage, and also points out common programming mistakes, such as dereferencing a possibly NULL pointer or running off the end of an array.

Given the rather trivial code snippet for `doit()` above, KLEE unsurprisingly generates two test cases, one with $x = 0$ and one with $x = 2147483905$. As KLEE's search can be nondeterministic, different values might be generated were it to be executed again. These two test cases exercise both of the paths. A slightly larger example is given in the section about coverage on page 256.

KLEE is particularly powerful because it includes a C library and can therefore handle calls to the normal C library functions: file handling, memory allocation, etc., including generating the content of files that are read by the program under test.

One serious danger that I have found with KLEE is that it symbolically executes the program under test (this is admittedly what it is designed to do!), and when searching, for example, for useful input to test a file deletion routine, it will actually delete files; when looking for inputs to test the `sleep()` library function, it may well execute `sleep(123456)` and thereby appear to lock up.

---

* `https://klee.github.io/`

** `llvm` is a popular compiler infrastructure: see `http://llvm.org/`.

Anecdote 25 *Some years ago when I first met KLEE, I wrote a really bad piece of code on which to exercise this, to me, new tool. I wrote a function that took two parameters: an integer* `length` *and a character string. I wrote the code in a convoluted manner to try to trick KLEE, and deliberately introduced a fault. If the* `length` *was greater than 151 and a multiple of 19, and if the last character in the string was the letter* `g` *the penultimate character was not a* `g` *and the letter* `g` *appeared exactly four times in the string, then my routine read one character off the end of the string array.*

KLEE found the problem and generated an error report and a test case with `length` *= 118891702 (which is greater than 151 and a multiple of 19) and the string = "xxxxggxgxxxg". This is the result I hoped to get. More unexpectedly, KLEE also found an additional fault. For certain values of the string, my code also read one character off the front of the array. I was sold!*

Summary

Static analysis has received bad press in the past because of the number of false and true but pedantic positives that the programmer has had to wade through. As reference [1] points out, even a 5% false positive rate running on a few hundred thousand lines of code can generate enough warnings to absorb several engineer-years of analysis.

Also, the results of the experiments described in reference [3] would seem to indicate that syntactic-level analysis is actually of little practical value.

However, with dynamic analysis becoming less useful, we have little choice but to place more emphasis on static analysis. Luckily, the improved algorithms are providing us with more sophisticated tools that can provide extremely useful analysis without too much investment of engineering time.

References

1. B. Blanchet, P. Cousot, R. Cousot, J. Feret, L. Mauborgne, A. Miné, D. Monniaux, and X. Rival, "A Static Analyzer for Large Safety-Critical Software," in *Proceedings of the ACM SIGPLAN 2003 Conference on Programming Language Design and Implementation (PLDI'03)*, (San Diego, California, USA), pp. 196–207, ACM Press, June 7–14 2003.
2. B. Livshits, M. Sridharan, Y. Smaragdakis, O. Lhoták, J. N. Amaral, B.-Y. E. Chang, S. Z. Guyer, U. P. Khedker, A. Møller, and D. Vardoulakis, "In Defense of Soundiness: A Manifesto," *Commun. ACM*, vol. 58, pp. 44–46, Jan. 2015.
3. S. Kim and M. D. Ernst, "Which warnings should I fix first?," in *ESEC/FSE 2007: Proceedings of the 11th European Software Engineering Conference and the 15th ACM SIGSOFT Symposium on the Foundations of Software Engineering*, (Dubrovnik, Croatia), pp. 45–54, September 5–7, 2007.
4. A. Mockus and D. M. Weiss, "Predicting risk of software changes," *Bell Labs Technical Journal*, vol. 5, pp. 169–180, 2000.
5. T. J. Ostrand and E. J. Weyuker, "How to measure success of fault prediction models," in *Fourth international workshop on Software quality assurance: in conjunction with the 6th ESEC/FSE joint meeting*, SOQUA '07, (New York, NY, USA), pp. 25–30, ACM, 2007.
6. R. Ceballos, R. M. Gasca, and D. Borrego, "Constraint Satisfaction Techniques for Diagnosing Errors in Design by Contract Software," in *Proceedings of the 2005 conference on specification and verification of component-based systems*, SAVCBS '05, (New York, NY, USA), ACM, 2005.
7. D. Crocker, "Can C++ Be Made as Safe as SPARK?," in *High Integrity Language Technologies Conference*, 2014.
8. M. D. Ernst, J. H. Perkins, P. J. Guo, S. McCamant, C. Pacheco, M. S. Tschantz, and C. Xiao, "The Daikon system for dynamic detection of likely invariants," *Science of Computer Programming*, vol. 69, pp. 35–45, Dec. 2007.
9. M. D. Ernst, J. Cockrell, W. G. Griswold, and D. Notkin, "Dynamically Discovering Likely Program Invariants to Support Program Evolution," *IEEE Transactions on Software Engineering*, vol. 27, pp. 99–123, Feb. 2001. A previous version appeared in *ICSE '99, Proceedings of the 21st International Conference on Software Engineering*, pages 213–224, Los Angeles, CA, USA, May 19–21, 1999.
10. C. Cadar and K. Sen, "Symbolic Execution for Software Testing: Three decades later," *Commun. ACM*, vol. 56, pp. 82–90, Feb. 2013.

VERIFICATION
VI

Chapter 19

Integration Testing

Today a usual technique is to make a program and then to test it. But: program testing can be a very effective way to show the presence of bugs, but is hopelessly inadequate for showing their absence.

Edsger W. Dijkstra

This chapter discusses some of the techniques recommended in ISO 26262 and IEC 61508 for use during integration testing. Again, I have chosen techniques perhaps slightly less well-known than the others:

Fault injection testing.
: See below.

Back-to-back comparison between a model and code.
: See page 284. The problem of producing a good test oracle is also discussed in that section.

Requirements-based testing.
: See page 288. This is included not so much because of its unusual nature, but more because of the impact that ISO 29119 may have on this type of testing.

Anomaly detection during integration testing.
: Anomaly detection is discussed in detail in the context of a run-time system starting on page 98. On page 291, the application of anomaly detection to the results of integration testing is covered.

Fault Injection Testing

Why Inject Faults?

For the distinction between a fault, an error, and a failure, see page 16. This section deals with the deliberate introduction of *faults* into a system. This is something that aggravates many designers and programmers: "I deliberately create my design/code as well as possible. Why would I deliberately put faults into it?"

Why one would inject faults is the primary question, and the primary answer must be so that the reaction of the system to unexpected events can be checked. If a sensor fails and starts producing random data, does the system detect this, report and otherwise ignore those data, and handle the situation correctly? These types of errors are difficult to test, but are crucially important for the continued safe operation of the system.

Qantas Flight 72 from Singapore to Perth on 7th October 2008, suddenly pitched downward, recording an acceleration of -0.8g, before the pilots could recover control. 12 passengers and crew were seriously injured. The final report into the incident (reference [1]) records:

> *... the aircraft had three ADIRUs* to provide redundancy and fault tolerance. Using the median of three values for a parameter as the system input is a common and generally robust algorithm, and the A330/A340 EFCS** used this approach for most parameters. However, in order to address aerodynamic issues associated with the locations of the three AOA*** sensors, the FCPC**** based the system input on the average value of AOA 1 and AOA 2.*
> ...
> *The algorithm was generally very effective, and could deal with almost all possible situations involving incorrect AOA data being provided by one ADIRU. ...*
> *Nevertheless, the AOA algorithm could not effectively manage a scenario where there were multiple spikes such that one triggered a memorisation period and another was present 1.2 seconds later. The problem was that, if a 1.2-second memorisation period was triggered, the FCPCs accepted the next values of AOA 1 and AOA 2*

* Air Data Inertial Reference Units

** Electronic Flight Control System

*** Angle of Attack

**** Flight Control Primary Computer

> *after the end of the memorisation period as valid. In other words, the algorithm did not effectively handle the transition from the end of a memorisation period back to the normal operating mode when a second data spike was present.* (Section 5.2.1 of reference [1])

In summary, the triplicated sensors (the ADIRUs) gave conflicting information. The main processing system had been programmed to handle this in general, but the particular combination of conditions that occurred on this occasion had not been anticipated. It is interesting that the algorithm had been changed as a result of another incident:

> *The A330/A340 FCPC algorithm for processing AOA data was redesigned after a problem was found with the initial algorithm during flight testing that was conducted before the aircraft type was certified. The redesign unintentionally introduced the design limitation in the algorithm ....*

Reference [1] goes on to point out limitations in both the failure mode and effects (FMEA) and fault tree (FTA) analyses that had been performed (see page 67 of this book for a description of these techniques).

As well as using injected faults to check the system's response to unexpected events, fault injection testing can also be used to estimate the number of Heisenbugs remaining in the system.

If you want to know the number of fish in a large pond, one possible technique to count them is to catch, say, 20 fish at random, mark them in some way and throw them back. After the marked fish have been allowed to mix with their unmarked friends, another 30 are caught at random. If it is found that 9 of these are marked, then we can estimate that there are about $N = \frac{30 \times 20}{9} \approx 67$ fish in the pond.

Similarly, if we inject B bugs into our system, particularly of the type where source code is changed in a realistic way, and the integration testing finds T bugs, of which M are those which we injected, then the number of bugs remaining can be estimated at about $\frac{T \times B}{M}$. This relies, of course, on the injected bugs being representative of the bugs already in the code. This is the IEC 61508 recommendation for measuring the efficiency of software testing (Annex C.5.6 of part 7).

In summary, injecting faults into a system allows us to exercise code for anomaly detection and handling that would otherwise be very difficult to test, and gives us a rough estimate of the number of Heisenbugs remaining in our system.

Where are Faults Injected?

Most embedded systems have a common structure (see Figure 19.1). The processing element reads input from one or more sensors, which may themselves be intelligent, performs some form of computation, and activates some output transducers (lights, brakes, screen displays, etc.).

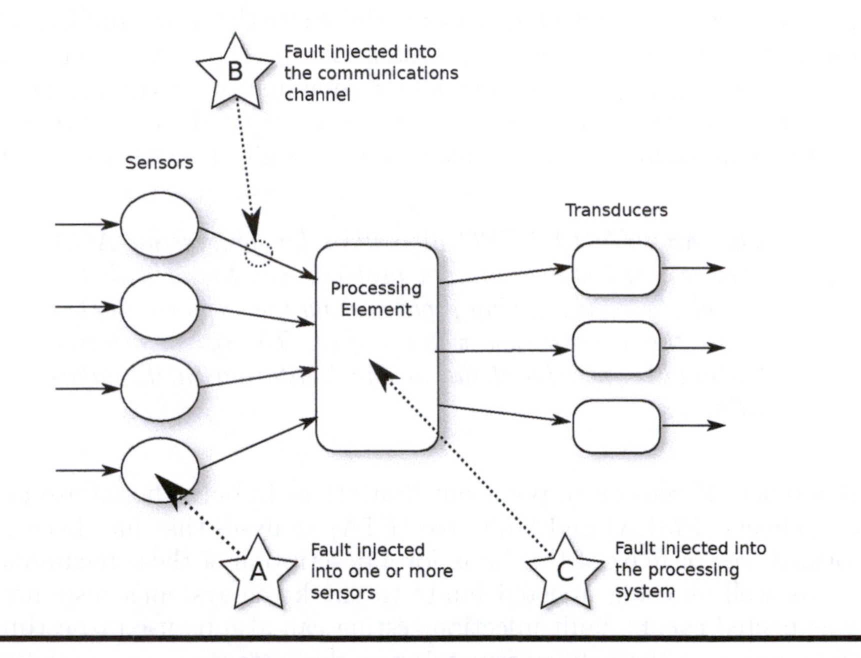

Figure 19.1 Fault injection points.

If a sensor misbehaves and starts passing false, misleading, or syntactically incorrect data to the processing element as happened on Qantas Flight 72, then this must be detected and handled. Generally, it is difficult to anticipate exactly what types of incorrect data might be received, and therefore also difficult to produce test cases for the processing element. The system's detection and handling of erroneous data can be examined by injecting faults into the sensor code (point A in Figure 19.1).

Injection point B is similar — if the data being sent by the sensor are corrupted, replicated, re-ordered, or lost, how does the processing element detect and handle this?

Point C is more subtle. Faults injected into the processing element itself are designed to invoke the types of error that would result from Heisenbugs and thereby test the resilience of the system: Are the errors resulting from the Heisenbugs detected and handled before they cause failures?

What Type of Faults are Injected?

The main difficulty associated with software fault injection is creating artificial faults that have the same characteristics as the (unknown) faults already in the design or code. Hardware fault injection, in contrast, is relatively straightforward — various bus or address lines are pulled low or high, or components are short-circuited at random intervals, and the behavior of the system noted.

Injected software faults are typically of one or more of the the following types.

Random changes to the source code of a program before it is compiled.
This could, for example, turn z = y + x into z = y - x to represent the type of error that a human programmer could have made.

Random changes to the object code of a compiled program.
Reference [2] by Henrique Madeira, *et al.* describes an experiment to see whether a software-implemented fault injection tool (SWIFT) can emulate the type of programming faults introduced by real programmers. One conclusion of the paper is that 44% of the faults introduced into programs by real programmers cannot be emulated by manipulating the object code of a program.

Corruption of static data.
For example, the linkages in a linked list could be corrupted, thereby simulating a race hazard* in the code that updates the list.

Corruption, replication, or discarding of messages.
This checks whether implicit assumptions have been made in the application about the reliable delivery of messages.

Manipulations of time.
For example, by artificially consuming processor cycles and preventing other programs from running.

Corruption of random memory locations.
This emulates a secondary event from a cosmic ray impact, to check the response of the system to hard and soft memory (DRAM) faults. The system's response to this type of fault is becoming increasingly important as the track widths within

* A race hazard is a Heisenbug, where very specific timing of two or more threads occasionally causes an error.

chips reduce and the chips become more susceptible to cosmic rays, crosstalk, electromagnetic interference, and thermal aging; see reference [3]. This form of corruption may alter a single bit, a byte, a word, a memory row, or a memory column.

Fault Injection and the Standards

IEC 61508 only recommends fault injection testing for hardware verification (see Annex B.6.10 of part 7) and to measure the efficiency of other software testing (Annex C.5.6 of part 7), whereas ISO 26262 recommends the technique for both software unit testing (Table 10 of part 6) and software integration testing (Table 13). Certainly, it is a technique that has been used with great success for software verification.

ADC, BCI, and Fault Injection

Our two imaginary companies (see Chapter 4) would probably carry out most of their fault injection testing in different ways.

For Alpha Device Corporation (ADC), which makes complete devices, insertion points A and B in Figure 19.1 are likely to be the most useful — how well does the system handle corrupt, missing, replicated, or misleading input from its sensors?

For Beta Component Incorporated's (BCI's) operating system, insertion point C is likely to be of primary usefulness — how quickly are corruptions to internal operating system data structures (whether caused by hardware or software errors) detected and handled?

Back-to-Back Comparison Test between Model and Code

The Standards

A model of the system, even in the form of a discrete event simulation (see page 14), can be an excellent source of test cases. The technique of performing back-to-back comparison between the model and the implementation is recommended in Tables 10 and 13 of ISO 26262-6 and in Table B.2 of IEC 61508-3, where it is called "Test case execution from model-based test case generation."

IEC 29119 also discusses model-based test case generation with approval in section 5.6.3 of part 1.

Back-to-Back Example

As an example, consider the BCI company (see Chapter 4), which is intending to certify its operating system. One operating system algorithm that superficially appears to be simple, but which has numerous nuances, is that used to avoid priority inversion. This can serve as an example of how a model may be used to derive test cases for the actual implementation.

The priority inversion problem relates to various tasks of different priorities needing to access the same resource, say a mutex. Assume that there are three tasks: T_1, T_2 and T_3 with priorities 1, 2 and 3 respectively (the higher the number, the greater the priority). The following sequence of actions on a single-core processor leads to the situation where T_3 is ready to run, but is blocked indefinitely by the lower priority T_2:

1. None of T_1, T_2 or T_3 is running.
2. T_1 becomes ready to run and seizes the mutex.
3. T_3 becomes ready to run, but cannot get the mutex because it is held by T_1 and so is suspended. This is intended behavior, and T_1 has been programmed so that it will only hold the mutex for a few microseconds.
4. T_2 becomes ready and, because it has a higher priority than T_1, pre-empts it. The highest priority task (T_3) is now blocked, waiting for T_1 to finish. But T_1 may never finish because it has been pre-empted by T_2.

It was a priority inversion bug of this kind that occurred to the Mars Pathfinder mission on 4th July 1997, when a medium-priority, long-running communications task pre-empted a low-priority meteorological data gathering task that was holding a mutex needed by the high priority information bus task.*

One mechanism for avoiding the priority inversion is known as priority inheritance. With this technique, the priority of the task holding the mutex is artificially raised to the highest of its natural priority and that of any task waiting for the mutex. In the example above, T_1 would have been raised to T_3's priority in step 3. This would have meant that it would have continued to run, even after T_2 came ready and the suspension of T_3 by a lower priority task would not have occurred.

* See, for example, `http://research.microsoft.com/en-us/um/people/mbj/Mars_Pathfinder/Mars_Pathfinder.html`.

The priority inheritance protocol seems very simple, but as BCI's designers have discovered, it contains many subtleties, particularly when priorities are lowered following the release of the mutex.

The designers take the decision to model their proposed algorithm using a discrete event simulation (known in IEC 61508 as "Monte-Carlo simulation" and recommended by ISO 26262-6, Table 3) and they carry out the following plan to check not only the algorithm proposed, but also to verify the implementation. The concept of discrete event simulation is described in this book starting on page 190. BCI's plan is to:

Write a discrete event simulation program to simulate the proposed algorithm under a long series of randomly generated conditions and check that priority inversion does not occur. Note that the SimPy simulation tool cannot be used for this because it doesn't allow priority inheritance!

Identify a number of distinct states in which each simulated thread may be (e.g., asleep, blocked on CPU, blocked on mutex by a lower priority thread ...)

Analyze the trace from the simulation to ensure that each of these states has been entered at least once for at least one thread, and a representative combination of thread states (including all states) have been reached system-wide. These steps raise the level of confidence that the algorithm is correct.

Generate combinatorial test cases (see page 257 and reference [4]) from the simulation traces — the actual C programs can be generated automatically. If, for example, the simulation trace says that thread 6 tries to seize mutex 4 at time $T = 15342$ and then gives up waiting at time $T = 15368$, the equivalent behavior can be built into the generated C program — it tries to seize the mutex at time $T = 15342$ with a timeout of $15368 - 15342 = 26$ time units.

Compile and execute the generated C programs. These are the test cases. They have been created from a model which is believed to represent a correct algorithm, and their selection by applying the combinatorial testing rules should ensure that the tests will adequately cover the implementation of the algorithm.

This is only one example of performing a back-to-back test between a model and code. Almost any (semi-)formal model can be used to generate good test cases.

Is Back-to-Back Testing Used?

Reference [5] by Robert Binder *et al.* provides the results of a survey carried out between mid-June and early August 2014 to see whether back-to-back testing (called model-based testing following the IEC 61508 terminology in the paper) is used in industry and, if so, whether companies using it are finding it effective. Unfortunately, the sample was quite small (100 responses, 86% of which were from industry, as opposed to academia and government, and 40% of which were from embedded software development) and self-selected, the invitation to participate having been circulated on social media such as LinkedIn.

Along with collecting various statistical data, the study asked several pointed questions about the quality of the testing, the time taken to learn the necessary techniques, during which phases of testing model-based testing was used, etc. The answers to these questions are summarized in reference [5] and more details are given on the associated website. The main conclusions are as follows:

- Over 60% of the respondents said that their (high) expectations of making test design more efficient were partially or fully met.
- 60% of the respondents said that the model-based testing was more effective (better tests) than whatever technique they had been using previously.
- Model-based testing was used primarily during system test, but was also applied during module, integration, and acceptance testing.
- The large majority of respondents said that they intended to continue using model-based testing.

The Model as a Test Oracle

There is one significant problem associated with the automatic generation of test cases: that of the so-called "test oracle." The oracle knows what the intended reaction of the system under test is to a particular input condition. It is relatively easy to use a test case generator, such as KLEE (page 273), to produce test cases providing good code and branch coverage. But, without access to the specification of what the program is intended to do, these tools cannot determine what the result of running the test case should be. It has to be left to the human programmer to perform the unenviable task of looking at each automatically generated test case and deciding what answer the system under test should return.

One enormous advantage of performing back-to-back verification against a model is that the model itself can act as the test oracle!

In the example of the priority inversion testing given in the previous section, the simulation trace provides the precise path through the state space that the system should follow. This ranges from "at the end of the test, thread 1 should be holding mutexes 4 and 6 with a priority of 23; thread 2 should be holding mutex 5 and waiting for mutex 4 with a priority of 15; ...." to a full trace of every event during the test.

In this case, there is no need for a human oracle — the model knows more about what should happen than any human could.

Requirements-Based Testing

Until recently, requirements-based testing was probably the conventional way of conducting a test — list the requirements that the module, component or system must satisfy and then prepare and execute test cases to demonstrate that it does satisfy those requirements. This is a recommended technique in ISO 26262 and IEC 61508.

What Are the Safety Requirements?

There are several problems with the requirements-based testing approach. The first is illustrated in Figure 19.2, which extends part of Figure 5.1 on page 55.

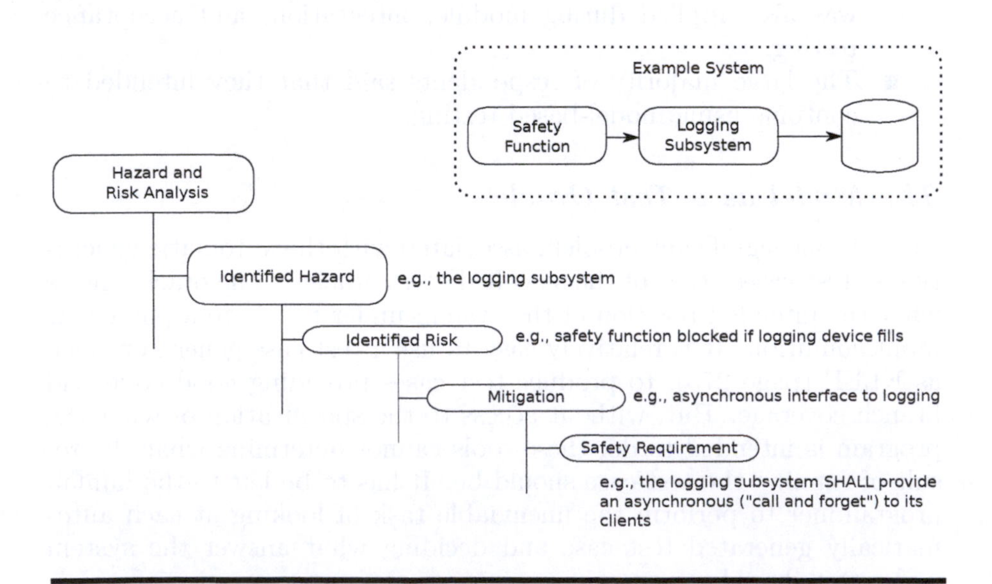

Figure 19.2 The origin of safety requirements.

Safety requirements are derived from risk mitigations. For example, assume that a device that provides functional safety is required to log its activities. The log is important, as it provides diagnostic information should a failure occur, but it is not part of a functional safety component of the system.

In this case, the logging subsystem would be identified as a hazard, and one risk associated with it might be that if the logging device (e.g., disk or nonvolatile memory) became full, the logging function would stop accepting information and the safety function would stop operating while waiting to log a message. One mitigation might be that the interface to the logging function be asynchronous ("call and forget") so that the safety function would not be suspended even if the logging subsystem stopped accepting input.

This would lead to a number of safety requirements, including the one shown in Figure 19.2: The interface to the logging subsystem must be asynchronous.

The question then arises of what the verification group should check once the device is in integration testing. If requirements-based testing is applied, then the answer is clear — each safety requirement is verified, either by inspection or by testing. In the case of the logging example, the verification group would check that the logging system provided an asynchronous interface, probably by inspection of the design or code.

However, I would argue that it would be much more useful for the verification group to look further back in the chain shown in Figure 19.2 and see what risk the safety requirements were mitigating. In this case the safety requirement was introduced to prevent the safety function being locked by a clogged logging subsystem. Tests could then be derived to establish that condition and check what happens. This not only confirms that the system remains safe, it also implicitly checks that the mitigations are adequate and have been correctly implemented.

Do We Know All the Requirements?

Another problem is that the systems we are building are becoming so complex that writing down the requirements, even the safety requirements, is becoming impossible. This is particularly true of accidental systems. Given the example of an accidental system on page 19, it can be argued that no set of requirements for the ship's radar would have led to tests under conditions of Global Positioning System (GPS) signals being jammed — largely because the system had evolved accidentally, and no one knew that it relied on GPS. By limiting testing to the verification of the requirements, one enormous risk has remained undetected.

Another factor that is now leading to the inadequacy of requirements is the incorporation of the human user into the system. Until a few years ago, the user was seen as a source of system failure, rather than being an integral part of the system itself. With the user on the outside, it was reasonably easy to define requirements for the system and test to those requirements.

More recently, advocates such as Nancy Leveson and Sidney Dekker have placed the human operator as an integral part of the system. As Dekker says in reference [6]:

> *Systems are not basically safe. People create safety while negotiating multiple system goals.*

Once we accept that our safety-critical systems are likely, at least in part, to be accidental and that unpredictable humans are a part of them, it becomes increasingly difficult to define a system's requirements, and therefore increasingly difficult to perform requirements-based testing.

ISO 29119 and Requirements-Based Testing

ISO 29119 ("Software Testing") recognizes this problem and, as it is a later standard than ISO 26262 (first edition) and IEC 61508 (second edition), it is not unreasonable to look to it for guidance:

> *The practicality of using requirements-based testing is highly-dependent on the context. . . . Even where requirements are well-specified, there is the danger that budgetary and time constraints may mean that it is not possible to test all requirements. When requirements are supplemented with information on their relative priorities then this can be used as a means of prioritizing the testing (in which case a risk-based approach can be used to prioritise the higher priority requirements). In practice, testers using requirements-based testing will often supplement the basic requirements in this manner so that the most important (highest risk) requirements are tested more thoroughly and earlier.*

The implication is that, in practice, requirements-based testing often becomes risk-based testing. Throughout ISO 29119, it is strongly implied that all testing is, in reality, risk-based and that this should be explicitly recognized and the risks properly assessed. Section 5.4 of ISO 29119-1 describes risk-based testing as follows:

> *It is impossible to test a software system exhaustively, thus testing is a sampling activity. A variety of testing concepts ... exist to aid in choosing an appropriate sample to test. ... A key premise of this standard is the idea of performing the optimal testing within the given constraints and context using a risk-based approach.*
> *This is achieved by identifying the relative value of different strategies for testing, in terms of the risks they mitigate for the stakeholders of the completed system, and for the stakeholders developing the system. Carrying out risk-based testing ensures that the risks with the highest priority are paid the highest attention during testing. Other standards can help with the determination of the risk of the systems in operation, e.g. ISO/IEC 16085 Risk Management.*

Given this guidance, it may be acceptable, when verifying an embedded product, particularly one with safety-critical implications, to recognize that requirements-based testing may be impractical, and to direct testing toward the areas of greatest safety risk.

Anomaly Detection During Integration Testing

Integration testing produces a vast amount of data, particularly if traces of events are gathered during the testing. Mehdi Zeinali *et al.* of the University of Waterloo, Ontario, are currently investigating theories, methods and tools to perform anomaly detection offline, in particular during regression* testing — see reference [7].

The idea is to analyze trace data produced by the system while it is under test to construct models of its "normal" behavior. These can then be compared to the system behavior recorded during subsequent integration tests. For example, if a pattern is found in the trace data for 30 successive tests over the course of several weeks, that indicates that a particular reply message is always sent within 10 milliseconds of

* Regression testing consists of repeatedly running a defined set of integration tests to ensure that changes made to the code to fix bugs or add features have not affected the basic behavior of the system.

the arrival of a particular stimulus, and this pattern changes in the 31st execution of the test, then investigation may be required. This change in behavior may, of course, be due to changes made in the structure of the software; it may, however, indicate that a change has occurred that accidentally, and incorrectly, changed system behavior.

This technique is based on the idea that most embedded systems have specific recurring behavior and thus the set of "normal" behaviors is naturally limited to a tractable size.

Discovered behavioral models can then also be used by online run-time monitors — see reference [8] by Samaneh Navabpour *et al.*

In the case of the two example companies introduced in Chapter 4, it would be particularly convenient for Beta Component Incorporated to supply its customer, Alpha Device Corporation (ADC), with a signature of the correct behavior of the operating system under load. ADC could then ensure that no anomalies occur in the operating system behavior during its testing of the complete device.

References

1. Australian Transport Safety Bureau, "Aviation Occurrence Investigation AO-2008-070," 2011. Available from `http://www.atsb.gov.au/publications/investigation_reports/2008/aair/ao-2008-070.aspx`.
2. H. Madeira, D. Costa, and M. Vieira, "On the Emulation of Software Faults by Software Fault Injection," in *In Proceedings of the International Conference on Dependable Systems and Networks*, pp. 417–426, IEEE Computer Society Press, 2000.
3. B. Schroeder, E. Pinheiro, and W.-D. Weber, "DRAM errors in the wild: A large-scale field study," in *Proceedings of the eleventh international joint conference on measurement and modeling of computer systems*, SIGMETRICS '09, (New York, NY, USA), pp. 193–204, ACM, 2009.
4. R. Kuhn, R. N. Kacker, and Y. Lei, *Introduction to Combinatorial Testing*. London: CRC Press, 2013.
5. R. V. Binder, B. Legeard, and A. Kramer, "Model-based Testing: Where Does It Stand?," *Commun. ACM*, vol. 58, pp. 52–56, Jan. 2015.
6. S. Dekker, *The Field Guide to Understanding Human Error*. Ashgate, Aldershot, Hants., rev. ed. ed., 2006.
7. M. M. Z. Zadeh, M. Salem, N. Kumar, G. Cutulenco, and S. Fischmeister, "SiPTA: Signal Processing for Trace-based Anomaly Detection," in *Proc. of the Conference on Embedded Software (EMSOFT)*, (New Dehli, India), Oct. 2014.
8. S. Navabpour, Y. Joshi, C. W. Wallace, S. Berkovich, R. Medhat, B. Bonakdarpour, and S. Fischmeister, "RiTHM: A Tool for Enabling Time-triggered Runtime Verification for C Programs," in *Proc. of the ACM Symposium on the Foundations of Software Engineering (FSE)*, (St. Petersburg, Russia), 2013.

Chapter 20

The Tool Chain

Validation of the Tool Chain

However good the design and however accurate its implementation, the integrity of a system relies on the correct operation of the tools used to create it. Perfect source code can be undone by a compiler that turns it into incorrect object code.

The standards that we are considering have much to say about how the tools we have chosen to create the final product must be analyzed, monitored, and justified. These tools range from editors used by programmers, through tools used for source code control (e.g., Apache Subversion or Git), system building (e.g., GNU Make and the compiler), problem reporting (e.g., Bugzilla), to the final tool that writes the software product onto the DVD for shipment. An unnoticed malfunction of any of these tools could lead to a substandard product being shipped.

Unfortunately, these types of errors are more than just theoretical possibilities. In reference [1], Xuejun Yang *et al.* make the interesting observation that:

> *Every compiler we tested was found to crash and also to silently generate wrong code when presented with valid input.*

This occurred not only on extremely complex code, but also when applying a "well-proven" version of the `gcc` compiler to a program as simple as that shown in Figure 20.1.

```
int foo (void)
    {
    signed char x = 1;
    unsigned char y = 255;
    return x > y;
    }
```

Figure 20.1 A superficially simple program.

Tool Classification

The incorrect behavior of some tools is less serious than the incorrect behavior of others. A fault that caused a text editor used by a programmer to display the variable `loopCounter` as `retnouCpool` would probably be immediately detected. A fault that caused a compiler to produce incorrect output code when the terminating value of a loop was 627 is less likely to be spotted and could be extremely dangerous for those few programs with such a loop.

Section 11 of ISO 26262-8 requires that all tools used in the development of the system be classified on two axes:

Tool Impact (TI).
: If the tool were to fail, could it introduce an error into the delivered system or fail to detect an error? If the answer to this question is "no," then a value TI1 is assigned to the tool, otherwise TI2.

Tool Error Detection (TD).
: If the tool were to introduce an error, is it likely that the error would be noticed? If detection is *very* likely, then a value TD1 is assigned; if detection is quite likely, then TD2; otherwise TD3.

Given these two values, a tool confidence level (TCL) can be assigned. The combination TI2/TD2 results in TCL2, and TI2/TD3 results in TCL3; any other combination results in TCL1. ISO 26262 then provides recommendations on how each tool should be analyzed and validated, the level of analysis depending on the TCL.

IEC 61508's classification is somewhat simpler. If the tool generates outputs that can directly or indirectly contribute to executable code, then it is classified as T3. If the tool is such that an error could fail to reveal a defect in the code, then it is T2. Otherwise, T1.

BCI's Tools Classification

Considering our fictitious company, Beta Component Incorporated (see chapter 4), the developers will identify a series of tools applied between the designers' brains and the shipped product. Each of these must be classified in accordance with ISO 26262 and IEC 61508.

In the light of the quotation from reference [1] above, it would be hard to argue that the compiler is not TI2/TD3 $\rightarrow$ TCL3 according to ISO 26262 and T3 according to IEC 61508.

BCI also uses Apache Subversion, a popular open-source repository, to maintain versions of its source code, test cases and documents. When a system is built, Subversion delivers copies of each source module as they were at a particular time. A bug in Subversion could deliver the wrong version of a module, resulting in an incorrect system being built. This tool is harder to classify. In accordance with ISO 26262, it must be TI2, but it is likely that the error would be noticed, particularly if version numbers were independently checked or the use of the wrong version led to a compilation or link error. Thus, this tool might be classified as TI2/TD2 $\rightarrow$ TCL2.

Note that by having version number checking defined as part of the build process, it may possible to reduce the TCL. In many cases, it may be economical to use process definition like this to reduce the amount of tool certification work.

Surprisingly, by IEC 61508's classification, Subversion could be T1, because it does not contribute directly to executable code (as a compiler would), and it does not fail to detect a code defect (as an incomplete set of test cases would).

Whatever the classifications that BCI proposes, they must be justified in the safety case.

For another example of the two example companies ascribing classifications to tools, see page 230.

Using Third-Party Tools

Most development tools are bought-in, rather than being produced within the company using them.

This creates particular problems because, in many cases of commercial tools, the source code is not available and, even if it were available, it would be an enormous exercise to understand and verify it.

If a tool has been suitably certified by the vendor, then this reduces, but does not eliminate, the work required by the user of the tool, who still has the responsibility of ensuring that it works correctly and that all bugs are analyzed for their impact on the project. But a certificate from a reputable certification company certainly helps in this, and such a certification would require the tool vendor to report bugs that are found and which could affect safety.

For open-source tools, perhaps the largest overheads for using the tool are validating it and assigning an engineer to keep track of the bug reports, looking for bugs that may impact its safe use. At the time of writing, there are 481 open bugs reported against the gcc C compiler. The assigned engineer would have to ensure that none of these bugs would impact the safety of the product being developed.

Even IEC 62304, generally quiet on technical topics, states:

> *If failure or unexpected results from SOUP* is [sic] a potential cause of the software item contributing to a hazardous situation, the manufacturer shall evaluate as a minimum any anomaly list published by the supplier of the SOUP item relevant to the version of the SOUP item used in the medical device to determine if any of the known anomalies result in a sequence of events that could result in a hazardous situation.*

Verifying the Compiler

Of all of the tools used to create the final product, the compiler poses the greatest challenge. Any compiler is an enormous piece of software dealing with very complex manipulation of what is often a poorly defined source language. In 2010, release 4.5.0 of the GNU gcc compiler consisted of more than 4 million lines of code and, at that time, it was growing at more than 10% per year. Verifying this type of tool to demonstrate that it always produces correct object code from *any* syntactically correct source file, and that it never produces code from *any* syntactically incorrect source file is an intimidating task.

I consider three approaches here:

1. The method of multiple compilers used in reference [1].
2. The method of generated programs as used by the National Physical Laboratory (NPL) in the UK.
3. The method of verifying the *compilation* rather than the *compiler*, as described in reference [2].

* Software of Unknown Provenance

Each of these techniques can be augmented by keeping good records of problems found in the compiler within the development organization. These problems are directly related to the specific use made of the compiler by that organization. Such evidence, particularly if it can be shown that no problem gave rise to the silent generation of incorrect code, may provide additional confidence in the compiler.

Multiple Compiler Approach

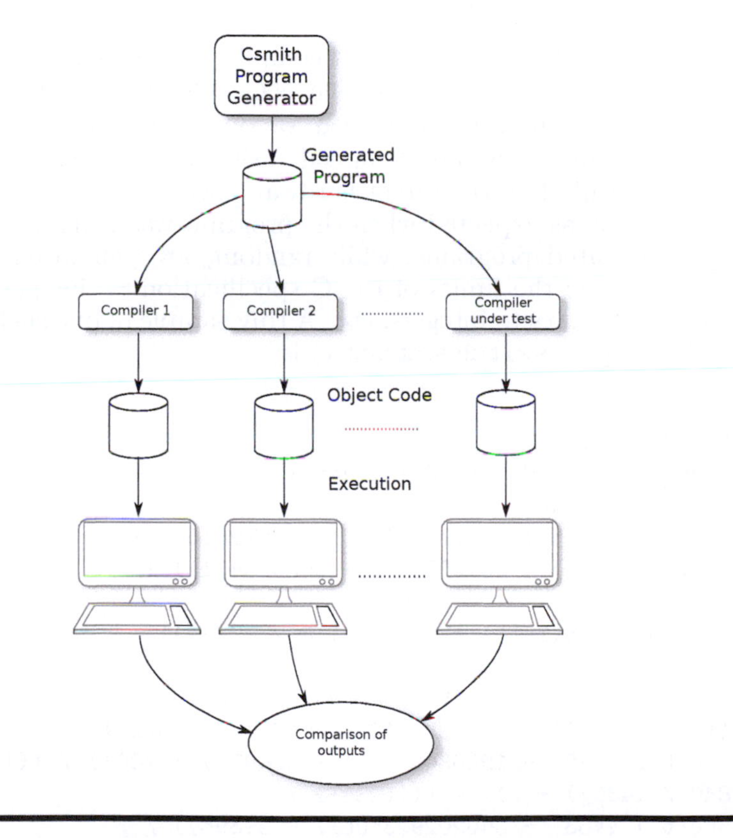

Figure 20.2 The multiple compiler approach.

Figure 20.2 illustrates the approach used in reference [1]. Random C programs were generated using the Csmith* tool. This tool generates C programs while avoiding the undefined and unspecified behavior of

* See `http://embed.cs.utah.edu/csmith/`

the C language.

Each generated C program was then presented to a number of different C compilers, and the resulting code was executed and the results compared. If all the compilers had been correct (or identically wrong!), then no differences would have been noted. In fact, as suggested by the quotation above, all tested compilers (including `gcc` and `llvm`) crashed on syntactically correct input and silently (i.e., without warning) generated incorrect object code.

The NPL Approach

Rather than apply several different compilers to the same source code, the NPL has produced a tool (known as CCST) that generates steered-random C programs with predefined, randomly-chosen, anticipated results. These programs are compiled, using the compiler under test, and then executed on the target hardware, see Figure 20.3. The result should then be as expected when the program was generated.

The generated programs, while random, are generated in such a way as to probe the limits of the C specification — integer overflow, promotion of `chars` to integers, etc. A tiny sample of one such program (a fairly complex `switch` statement) is:

```
{int index = 0;
for(index = 0; index < 685; index++) {
V63[index] = index;
}}V61 = (((((int ) (*V17)[0] ) - 1218795160) *
((((int ) - V3 ) + 162000423) + 590638431)) +
((((int ) - V35 ) - 14306004) /
(955746002 + (((int ) *V52 ) + 178622297))));

switch((((((((short ) (V16  - 7097)) + 3010) - 112) -
(((((short ) ((*V17)[0]  - 1218769428)) - 25727) *
(((short ) (*V8  - 1900348.0f)) - 22497)) + 280)) / ((1 *
(11840 / 1480)) + 0)) + ((((23472 /
(((short ) (V33  - 3602029235UL)) - 29396)) /
(((short ) - V7 ) + 1359)) *
(((short ) ((*V10)(V12)  - 139147259U)) -
14013)) + (((short ) (V16  - 20212)) - 16143)))) {

case (((30368 / 208) + (10 + 9)) * -2) + -18):
SETERROR;
break;
default:break;
}
```

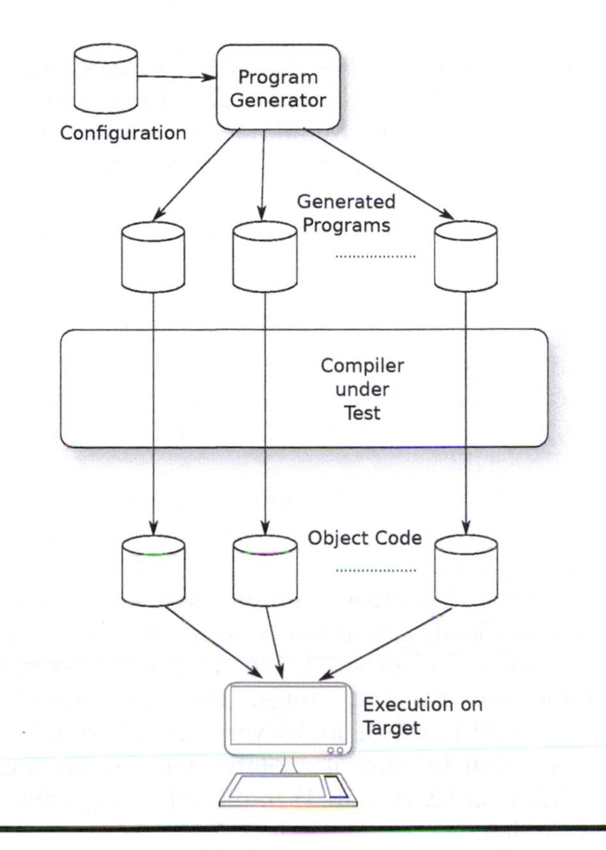

Figure 20.3 The NPL approach.

Those are a few lines from the middle of a 1500 line generated C program that, when preprocessed, compiled, assembled, linked, loaded and executed, should produce a particular value. This tool is particularly valuable because it exercises not only the syntax-checking front-end of the compiler, but also the code-generation back-end, the assembler, the linker and the loader.

I have used this tool to find several errors in the well-used `gcc` compiler.

The Compilation Validation Approach

Imagine that, during the design of a safety-critical product, you need to know an approximate square root of 17. You use a pocket calculator and it tells you that it is roughly 4.1231. Because this will be used as part of the design of a safety product, you would like to confirm that this is correct before you use it. You have two choices: either you can square 4.1231 (preferably on a different calculator) and check that it gives a value close to 17, or you can examine the algorithm used by the calculator and prove that it will *always* give the correct answer to a square root calculation. No one would consider the second option, but this is precisely what we do when trying to demonstrate the correctness of a compiler.

To verify a compiler, it is necessary to show that it generates correct code from *any* syntactically correct input and, more importantly, that it refuses to produce code from *any* syntactically incorrect input. This is an impossible task, and the question arises of whether such verification is actually needed. Frankly, if the compiler generates wrong code for every function containing a variable whose name begins, for example, with the letter `x`, that doesn't matter, as long as our code doesn't contain any such variables.

The bugs that the NPL tool found in the `gcc` compiler were in code constructions so complex that they would never have been written by a human programmer. The compiler bugs were genuine, but would never have caused an error in human-written code. As more and more code is automatically generated from models (see page 218), this statement may need to be revisited, but, at present, compiler testing is finding genuine, but irrelevant, bugs. Even the snippet of code illustrated in Figure 20.1, which looks very trivial, should never have passed code inspection because it violates most C coding standards (including MISRA-C 2012, rule 10.4) in directly comparing a signed to an unsigned variable.

Given that we are interested only in compiler bugs that affect our code, it should be possible to replace *compiler* verification with *compilation* verification — whether or not the compiler is always correct is irrelevant; it is only important to know whether it was correct for *our* particular compilations. In a same way, it is not important whether the calculator we used to find the square root of 17 is always correct, only whether it was correct in that one, specific, calculation.

Reference [3] provides not only a survey of compiler verification techniques, but also an approach to proving that a particular compilation is correct. The technique is to use "certificates" produced by the compiler under test to generate a mathematical proof that the compilation is correct. This needs to be handled very subtly, because the certificates themselves, being produced by an untrustworthy compiler, cannot be trusted either. Although this is almost certainly the path that will be followed in the future, the technique is not yet 100% practical.

References [2] and [4] by the present author and Marcus Bortel propose an alternative approach which, while not providing a proof of the correctness of the compilation, provide a level of confidence that can be chosen to match the resources available.

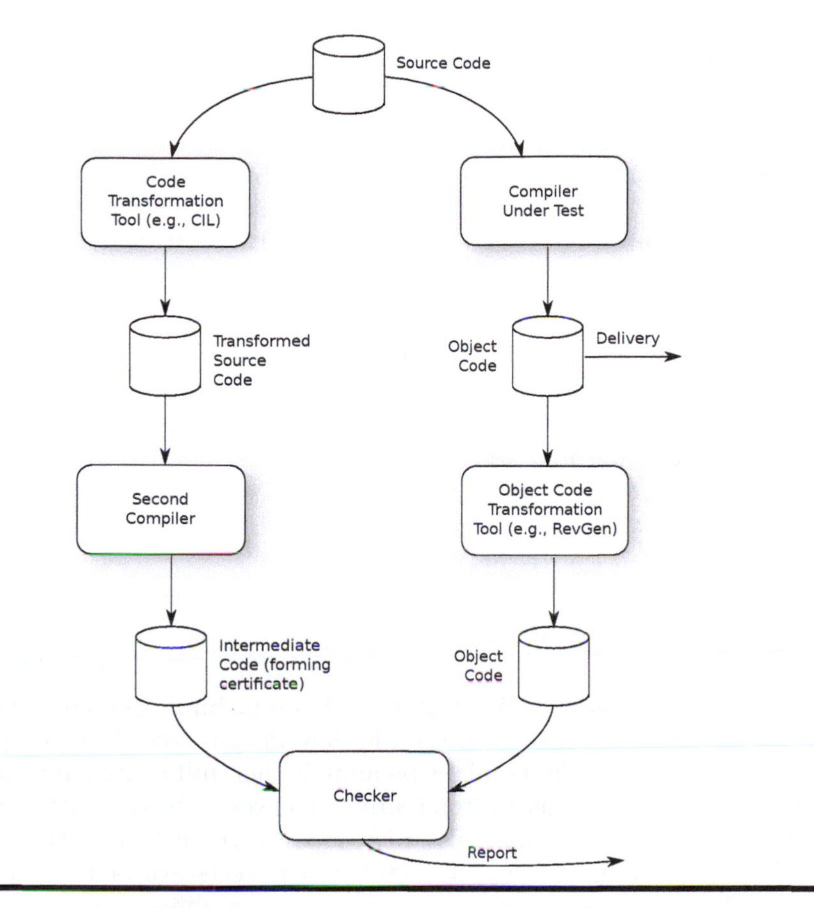

Figure 20.4 A pragmatic approach.

Figure 20.4 illustrates the technique. It uses two compilers, as in the technique illustrated in Figure 20.2, and these compilers may operate on syntactically different versions of the program (e.g., the standard version of the program and a version once it has been preprocessed by a tool such as CIL — see reference [5] by George C Necula *et al.*).

The output from the second compiler is used to generate invariants related to the data- or control-flow of the program and these invariants are applied against the output of the compiler under test. Any failures of the invariants indicate that one or the other compilers is faulty, and manual inspection is needed to determine which.

For example, given the snippet of code below, symbolic execution, for example with KLEE, can derive the two invariants that must hold at the end of the execution:

1. $\forall x(0 \leq x < i) \rightarrow a[x] \leq max$
2. $i \geq len$

The second of these does not relate to an observable variable, but the first does and so should be true in both copies of the object code. This can be automatically checked, providing increased confidence in the compilation.

```
int findMax(int *a, int len) {
    int max = a[0];
    int i = 0;
    for (i = 0; i < len; i++) {
        if (a[i] > max)
            max = a[i];
        }
    return max;
    }
```

This technique has the advantage that it can be implemented incrementally. Perhaps in phase 1 only basic flow graphs would be compared; later data flow invariants might be added; then full control flow invariants. It may also be useful to create test cases from the output of the second compiler (e.g., using KLEE) and use these to test the output of the compiler under test. Effectively, the tests are being generated from what should be equivalent code, but any differences introduced by the compiler might lead to test cases that cannot be satisfied by the output of the compiler under test.

The disadvantages of this technique include the fact that it doesn't provide a *proof*, only a level of confidence, and that it is difficult to apply at high levels of compiler optimization, where control and data flows may be significantly changed. However, high levels of compiler optimization may not be suitable for a safety-critical application: sections 7.4.4.4 of IEC 61508-3 and 11.4.4.2 of ISO 26262-8 include warnings about optimizing compilers.

ADC's and BCI's Compiler Verification

Our two fictitious companies (see Chapter 4) have slightly different problems.

Beta Component Incorporated (BCI) is producing low-level operating system code, primarily using the C language. Reference [1] mentions such code explicitly:

> *We claim that Csmith is an effective bug-finding tool in part because it generates tests that explore atypical combinations of C language features. Atypical code is not unimportant code, however; it is simply underrepresented in fixed compiler test suites. Developers who stray outside the well-tested paths that represent a compiler's "comfort zone" — for example by writing kernel code or embedded systems code, using esoteric compiler options, or automatically generating code — can encounter bugs quite frequently.*

In this case, BCI may wish to use the NPL tool (now archived by the IEC61508 Association)* to build a suitable level of confidence in its compiler. However, as the quotation above points out, kernel code is atypical — it contains, for example, much more pointer manipulation and many more bit operations than would be expected in application code, so it may be necessary to extend the ideas underlying the NPL tool to incorporate more such constructions.

Alpha Device Corporation's (ADC's) problems are, if anything, worse. ADC is writing higher-level application code that will run on BCI's operating system. It is proposed that this higher-level code be written in C++. A C compiler is extremely difficult to verify; a C++ compiler is much worse because of the additional features of the C++ language. Of course, the C++ compiler should pass all the C compiler tests, but that is hardly scratching the surface of the C++ features. The C++ library modules would also have to be verified, including the standard template library if ADC uses this.

I have been unable to find an equivalent of the NPL program or Csmith for C++. It would not be impossible to extend Csmith to generate C++, but, even for a small subset of C++, as would be permitted under ADC's coding standard, this would be a large amount of work that does not contribute directly to creating ADC's device.

One useful attribute of the approach described in reference [4] is that it is agnostic to the source language — it is irrelevant whether the code is written in C, C++ or even FORTRAN. This may be a possible approach for ADC.

To build an argument for the safety case justifying the use of C++, ADC really has three options:

---

* `http://www.61508.org/`.

To demonstrate the correctness of the C++ compiler and libraries. As described above, this is likely to be difficult, even if only a small subset of C++ is permitted in ADC's coding standard. The history of bugs found over the previous years in the C++ compiler within ADC could provide additional evidence of the robustness of the compiler.

To argue that postcompilation verification is sufficient to detect any erroneous code generated by the compiler. This argument says that the compiler is really an extension of the (human) programmer. The human programmer may make mistakes, and the purpose of verification (dynamic testing, etc.) is to catch these. The compiler simply takes the programmer's code and converts it into another language, warts and all. If the compiler adds a few more mistakes, then these are probably no more than the programmer would have made had the code been originally written in assembler. Given the level of verification that will be performed, any errors introduced by the compiler are irrelevant.

This is an interesting argument, undermined by the limited number of system states that can be covered during dynamic testing. However, using the "risk-based" testing technique described in ISO 29119, it might be possible to focus testing on aspects of the system where compiler error may have produced misoperation. This would not be an easy argument to justify.

To argue that the system architecture is such that incorrectly generated code would not affect the safety of the device. Using a technique such as a diverse monitor — see page 147 — it might be possible to design the system in such a way that any potentially unsafe activity by the compiled application would be caught and rendered harmless by the monitor. This would probably increase ADC's product cost, but it would reduce the certification costs.

Given these hurdles, it might even be advantageous for ADC to reconsider its choice of programming language. If ADC has legacy code written in C++ that is to be included in the new device, it might be worth considering the possibility of isolating it during the design phase so that it cannot affect other code.

References

1. X. Yang, Y. Chen, E. Eide, and J. Regehr, "Finding and Understanding Bugs in C Compilers," in *Proceedings of the 2011 ACM SIGPLAN Conference on Programming Language Design and Implementation*, ACM SIGPLAN, ACM, June 2011.
2. C. Hobbs, "Compiler or Compilation Validation?," in *2014 Safety Critical Systems Symposium*, SSS '14, (Brighton, UK), Safety-Critical Systems Club, 2014.
3. J. O. Blech and B. Grégoire, "Certifying compilers using higher-order theorem provers as certificate checkers," *Formal Methods in System Design*, vol. 38, no. 1, pp. 33–61, 2010.
4. C. Hobbs and M. Bortel, "Compiler oder Compilierungs-Validation?," in *Proceedings of ESE Kongress 2013*, 2013.
5. G. C. Necula, S. McPeak, S. Rahul, and W. Weimer, "CIL: Intermediate Language and Tools for Analysis and Transformation of C Programs," in *Proceedings of Conference on Compiler Construction*, Conference on Compiler Construction, 2002.

Chapter 21

Conclusion

I hope that you have found some stimulating thoughts in this book.

I know that some of the points I have made are controversial, and we may have to agree to disagree on whether software failures fundamentally differ from hardware failures and cannot be treated statistically, whether code coverage metrics are, in themselves, a useful goal during module testing, whether the goal structuring notation is better for a safety case argument than Bayesian belief networks, and even whether `emacs` is a better editor than `vi`.

I have one further recommendation: that you obtain a copy of *Software for Dependable Systems,* published by the National Research Council.* This is available both as a published book (ISBN 978-0-309-10394-7) and as a free `pdf` download. It is somewhat repetitive, but it describes what I agree are the three keys to a successful and safe software development. These are the "three Es":

Explicit Claims. No system is dependable under all circumstances, and it is important that we, as developers, and our customers are aware of precisely what we are, and are not, claiming.

Evidence. Concrete evidence must be available to support our safety argument.

Expertise. Building a safe system is difficult, and developers need expertise in software development, in the domain where the product will operate (medical, industrial, automotive, railway) and in the broader safe-system context.

* Nowhere in the book does it mention in which nation the Research Council is located. Judging by the authors, it would seem to be the UK or the USA.

As you put this book aside and return to software development, ensure that your claims are explicit, that your evidence is solid, and that you maintain your expertise by following practical research being published in the journals and at conferences.

APPENDICES VII

Appendix A

Goal Structuring Notation

Background

The description of the contents of the safety case starting on page 61 includes justification for expressing the safety case argument in a semi-formal notation. There are various notations that might be used; this appendix gives a brief description of the goal structuring notation (GSN) and Appendix B describes Bayesian belief networks (BBNs)

The GSN* is a graphical notation used to "document explicitly the elements and structure of an argument and the argument's relationship to evidence." It is standardized in reference [2].

Paragraph 5.3.1 of ISO 26262-10 and reference [3], prepared with the co-operation of the USA's Food and Drug Administration, explicitly mention the GSN as a suitable notation for a safety case.

In summary, the graphical symbols are as follows:

- A rectangle contains a claim (known as a "goal" in GSN-speak). This will be the statement whose truth we want to demonstrate. *For example: "The German word for Friday is Freitag."*
- A parallelogram contains a strategy: What strategy is used to justify the claim?
  For example: "The strategy to determine whether the German word for Friday is Freitag will be to consult a reputable dictionary."
- A rounded rectangle contains a context: The claim is made

* See http://www.goalstructuringnotation.info/ and document [1].

within a particular context.
For example: "We only claim that German word for Friday is Freitag in modern German — we make no claim about older forms of the language."

- An oval contains an assumption: The strategy is acceptable if it is assumed that ...
For example: "We assume that there is a single German word for Friday (unlike, for example, Saturday)."
- A circle represents a solution: effectively a description of the evidence that will be presented to justify the claim.
For example: "The evidence to support the belief that the German word for Friday is Freitag is the entry on page 234, second column, of Mrs. Merkel's EU wordlist (ISBN 12-120192-199281)."
- A diamond represents incomplete work; there should be none of these in the final case.

Example

To demonstrate the GSN format, consider the simple example of justifying our claim that a particular algorithm that we will be using in a device is correct. We have the following pieces of evidence:

- A formal proof of a simplified version of the algorithm, the full algorithm having proven intractable for the formal proving tool. The simplified algorithm has all the characteristics of the real algorithm apart from the use of infinite-precision numbers rather than floating point ones.
- A discrete event simulation of the entire algorithm, exercising not only the normal input conditions, but also many corner cases.
- Records of a code inspection carried out on the implementation of the algorithm.
- Results from our test laboratory, where a large number of dynamic tests have been run on the implementation of the algorithm. Where possible, these have been checked against the simulation results, but it has not been possible to reproduce in the laboratory all the conditions that were met in the simulation (this is the normal relationship between simulation and dynamic testing — the simulation can exercise more corner-cases).

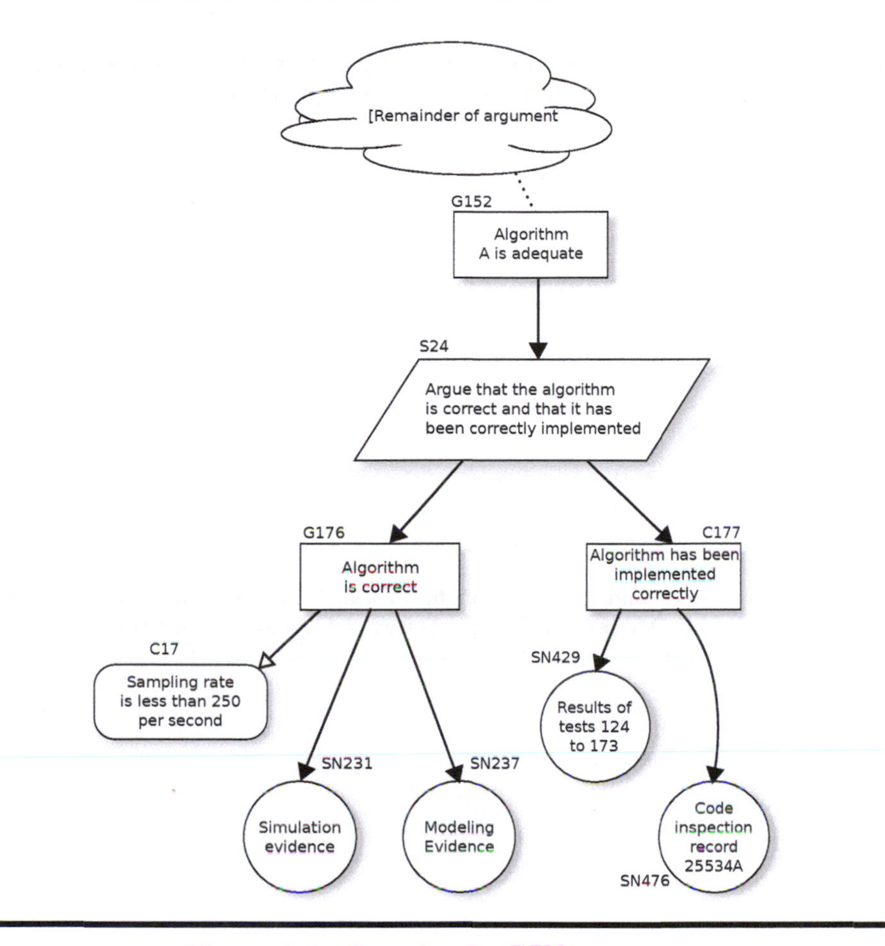

Figure A.1 Sample of a GSN argument.

Figure A.1 illustrates this argument in GSN form. This would normally be a tiny part of a much larger picture. The main claim (G152) is that the particular algorithm is adequate. Our strategy is to break this claim into two and show that the algorithm is correct and that our implementation of it is correct. We then provide links (SN231, SN237, SN429, SN476) to the evidence that we believe justifies each of those claims. Note that we only make the claim for the correctness of the algorithm in the context of a sampling rate less than 250 samples per second (C17). Each of the evidence identifiers should point directly to the document, analysis, or file of test cases and results.

It is this type of diagram that could be taken to the auditor early in the development process, asking the question: "If we were to present the evidence to support this argument, would it convince you?" The auditor might in this case reply that additional evidence in the form of an analysis of the floating point rounding errors would be required — something it is better to find out earlier rather than later.

GSN or BBN?

A brief comparison of using the GSN and the BBN notations for expressing a safety case argument is given on page 324.

References

1. T. Kelly and R. Weaver, "The Goal Structuring Notation — A Safety Argument Notation," in *Proc. of Dependable Systems and Networks 2004 Workshop on Assurance Cases*, 2004.
2. G. S. Committee, "GSN Community Standard Version 1," 2011.
3. C. Weinstock and J. Goodenough, "Towards an Assurance Case Practice for Medical Devices," tech. rep., Software Engineering Institute, Carnegie Mellon, 2009.

Appendix B

Bayesian Belief Networks

Bayesian Belief Networks (BBNs) are useful both for expressing the argument of a Safety Case and for encoding a fault tree.

BBNs are described in many publications, for example, references [1] by Bev Littlewood and Lorenzo Strigini, and reference [2] by Norman Fenton and Martin Neil. This latter reference provides examples of BBNs in both the contexts used in this book: to represent the argument in a safety case and to represent a fault tree in a failure analysis.

For a very straightforward introduction to the whole area of Bayesian statistics, see reference [3] by Mary Cowles.

Frequentists and Bayesians

Although it is not strictly necessary to understand the difference between frequentists and Bayesians in order to use the tools needed to build a BBN, such understanding might help you understand some of the Bayesian concepts, such as prior probabilities, better.

Consider the statement, "If I toss a fair coin, the chances of it landing heads uppermost is 0.5." This is fairly uncontroversial, but what does it mean? A frequentist would say that it means that, were you to toss the coin 100 times, it would be surprising were the number of heads very different from $100 \times 0.5 = 50$.

There are two problems with this explanation. The first lies in what is meant by "very different"? If I did the experiment and got 47 heads, would that be a surprise? How much of a surprise? The frequentist might say that the probability of 47 of heads occurring is 0.067, but that doesn't help because it's defining probability in terms of probability.

The second problem is more profound. What about events that cannot occur many times? What does the statement "There is a 50% chance that there will be a thunderstorm tomorrow" mean? It isn't possible to have 100 tomorrows and check that a thunderstorm occurred on about 50 of them.

With embedded devices we may be dealing with probabilities of failure per hour of use so low that it would be impossible to collect enough data to make a frequentist argument.

In contrast to the frequentist approach, the Bayesian thinks in terms of strength of belief: how strongly do you believe that there will be a thunderstorm tomorrow? Of course, if you're a meteorologist and have access to today's weather data, you might have a fairly sharply defined belief (e.g., 35% confident that there will be a thunderstorm, 65% confident that there won't). If you are not a meteorologist, then you could perhaps do no better than select "non-informative" values of 50%/50% or assume that, at this time of the year, thunderstorms occur once every ten days and assume that the probability is 10%/90%.

Prior Probabilities

The meteorologist and non-meteorologist, specifying their strength of belief in a thunderstorm occurring tomorrow, are estimating "priors," "prior" here meaning "before looking at the data". Similarly, an experienced auditor or new developer arriving at our fictitious company, Beta Component Incorporated (see Chapter 4), would be able to assign prior probabilities to the thoroughness of, for example, the company's fault-injection testing. We can assume that BCI is a long-established company that has been providing components into safety-critical devices for some time. The assigned prior probabilities then might be:

Highly confident of finding thorough fault-injection testing	5%
Confident of finding thorough fault-injection testing	70%
Neutral about finding thorough fault-injection testing	20%
Doubtful of finding thorough fault-injection testing	5%
Highly doubtful of finding thorough fault-injection testing	0%

These are illustrated graphically in Figure B.1. There is no restriction on the manner in which the prior judgement of the expert is captured. For example, Figure B.2 shows a continuous form of the same belief (actually, a β-distribution with parameters 10 and 40). In the case of experienced auditors, their levels of belief will presumably be based on

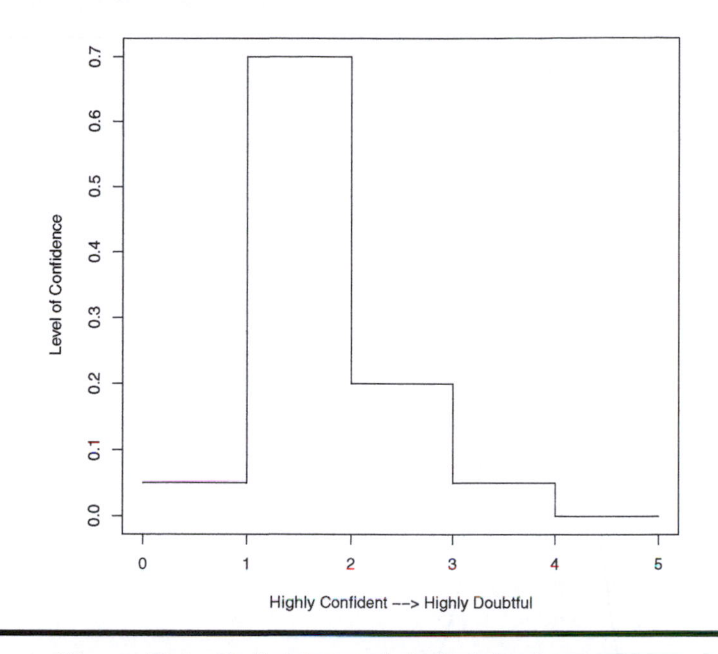

Figure B.1 Belief in good fault-injection testing.

having previously inspected dozens of similar companies.

Reference [3] describes in more detail how knowledge from experts might be captured in these forms.

Once the prior probabilities are assigned, they are fed, together with the actual observations, into Bayes' theorem.

Bayes' Theorem

The Rev. Thomas Bayes' famous theorem (reference [4]) was actually published a couple of years after his death, having been augmented by Richard Price. The theorem is as simple as it is powerful:

$$P(A|B) = \frac{P(B|A) \times P(A)}{P(B)} \quad \text{if} \quad P(B) \neq 0 \tag{B.1}$$

$P(A)$ means the probability of A occurring, and $P(A|B)$ means the probability of A occurring given that B has occurred.

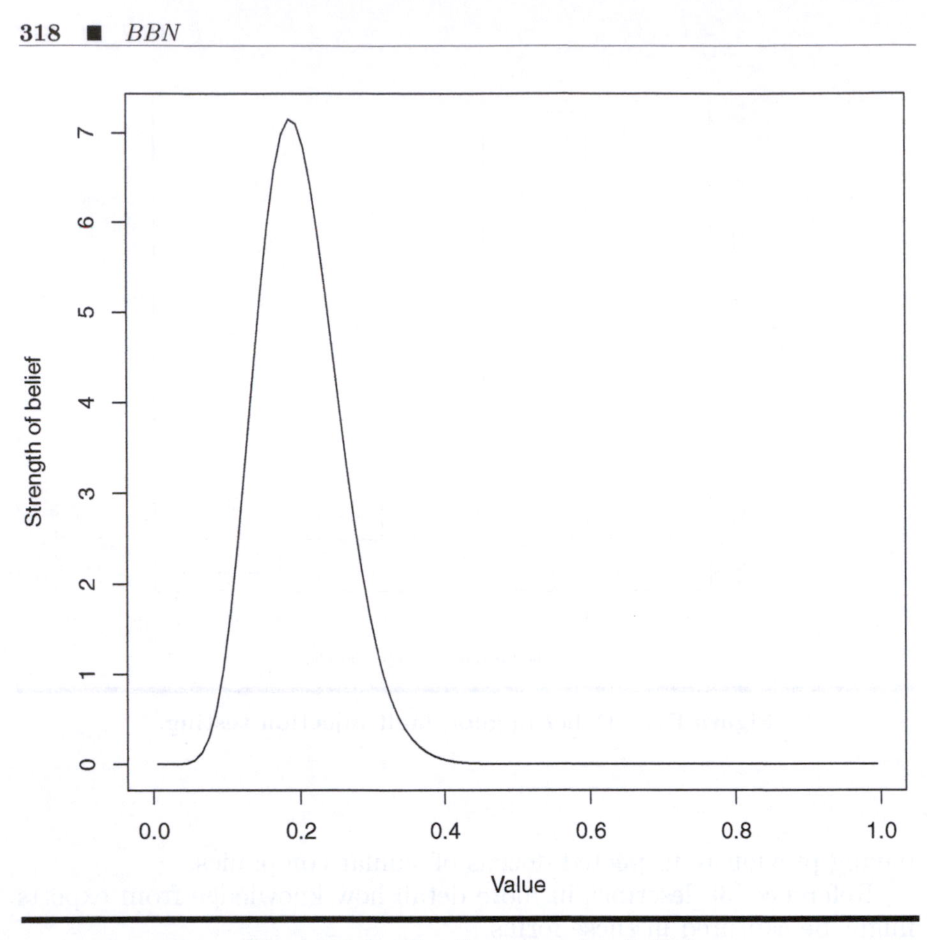

Figure B.2 Continuous prior belief.

A Bayesian Example

A particular disease affects 0.01% of the population. There is a test for the disease that always indicates that the patient has it, if she does (that is, it does not indicate false negatives). Unfortunately, it has a 5% false positive rate: 5% of the time it indicates that the patient has the disease when she hasn't.

Ethel walks into a clinic on impulse and takes the test. She tests positive. What is the probability that she has the disease?

We can let: A mean "Ethel has the disease," and B mean "Ethel has a positive test." We would like to calculate $P(A|B)$: the probability that Ethel has the disease, given that she has a positive test.

In Equation B.1 we need values for:

- $P(B|A)$. That is, the probability that Ethel will have a positive test given that she has the disease. Clearly, $P(B|A) = 1$, because the test never gives a false negative.
- $P(A)$. This is the prior probability that Ethel has the disease before we take the test into account. As 0.01% of the population has the disease, $P(A) = 0.0001$.
- $P(B)$. This is the prior probability of getting a positive test. One in 10,000 people has the disease and will therefore certainly get a positive test, the other 9,999 people out of 10,000 do not have the disease, and they have a 5% chance of getting a positive result. So
$P(B) = 1 \times \frac{1}{10000} + 0.05 \times \frac{9999}{10000} = 0.050095$

Given these values we can apply equation B.1 to calculate:

$$P(A|B) = \frac{P(B|A) \times P(A)}{P(B)} = \frac{1 \times 0.0001}{0.050095} = 0.001996 \qquad \text{(B.2)}$$

So, given Ethel's positive test result, there is only an approximately 0.2% chance that she has the disease. This result comes as a surprise to many people (including doctors!), because in some sense the test is "95% accurate." But the justification is easily seen. Consider a group of 10,000 people. On average 1 will have the disease and, of the remaining 9,999, about 500 will get a positive test result although they don't have the disease. So, of the 501 people with a positive result, only 1 (about 0.2%) will have the disease.

It is hardly worth bringing a tool to bear on this simple example, but I have done so, see Figure B.3. The drawing to the left of that figure indicates the situation where no information is available about Ethel's test result, it just assumes the prior probabilities; the drawing on the right gives the probabilities when there is a positive test result. As expected, with the test being 100% positive, the chances that Ethel has the disease is about 0.2%. Note the "backwards" reasoning: from the effect (a positive test result) to the cause (Ethel having the disease).

What Do the Arrows Mean in a BBN?

In Figure B.3 arrows are drawn from the "has disease" to the "test is positive" boxes, indicating a causal relationship: if the person has the

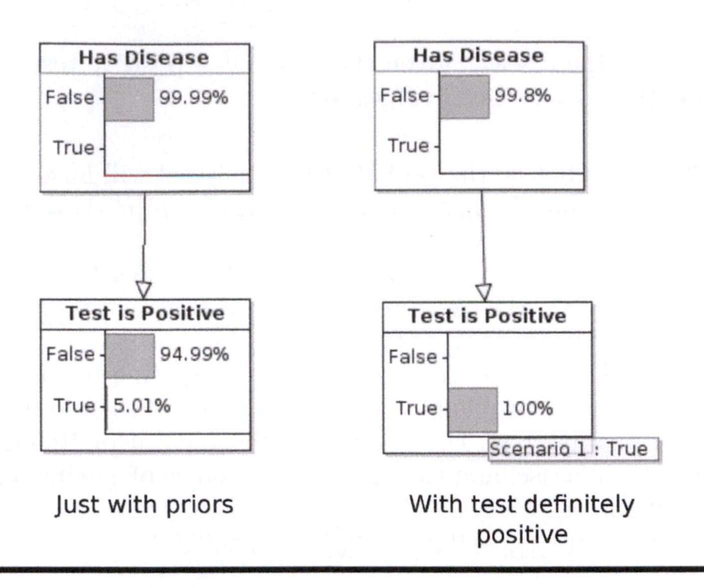

Figure B.3 Ethel's test results.

disease, it will affect whether the test is positive. It would have been just as easy to draw the arrows in the opposite direction, in which case they could be interpreted as meaning "adds weight to the belief that." So a positive test adds weight to the belief that the person has the disease.

As long as the usage is consistent, the actual meaning of the arrows is not that important. Reference [5] by Martin Lloyd and the present author provides a number of possible meanings for the arrows in the context of a safety case and describes different Bayesian "idioms": definitional, extended evidence, process-product, measurement, and the induction idioms.

BBNs in Safety Case Arguments

BBNs are very useful for specifying (and quantifying) the argument within a safety case. As an example, consider Figure B.4.

This figure illustrates what might be a very small part of a safety case. It presents the sub argument that the coding standard being used on the development project is effective, the same example as was used on page 65. The analyst intends to argue that, if the coding standard actually exists, if it is of high quality, and if the programmers are actually applying it in their work, then the coding standard is effective.

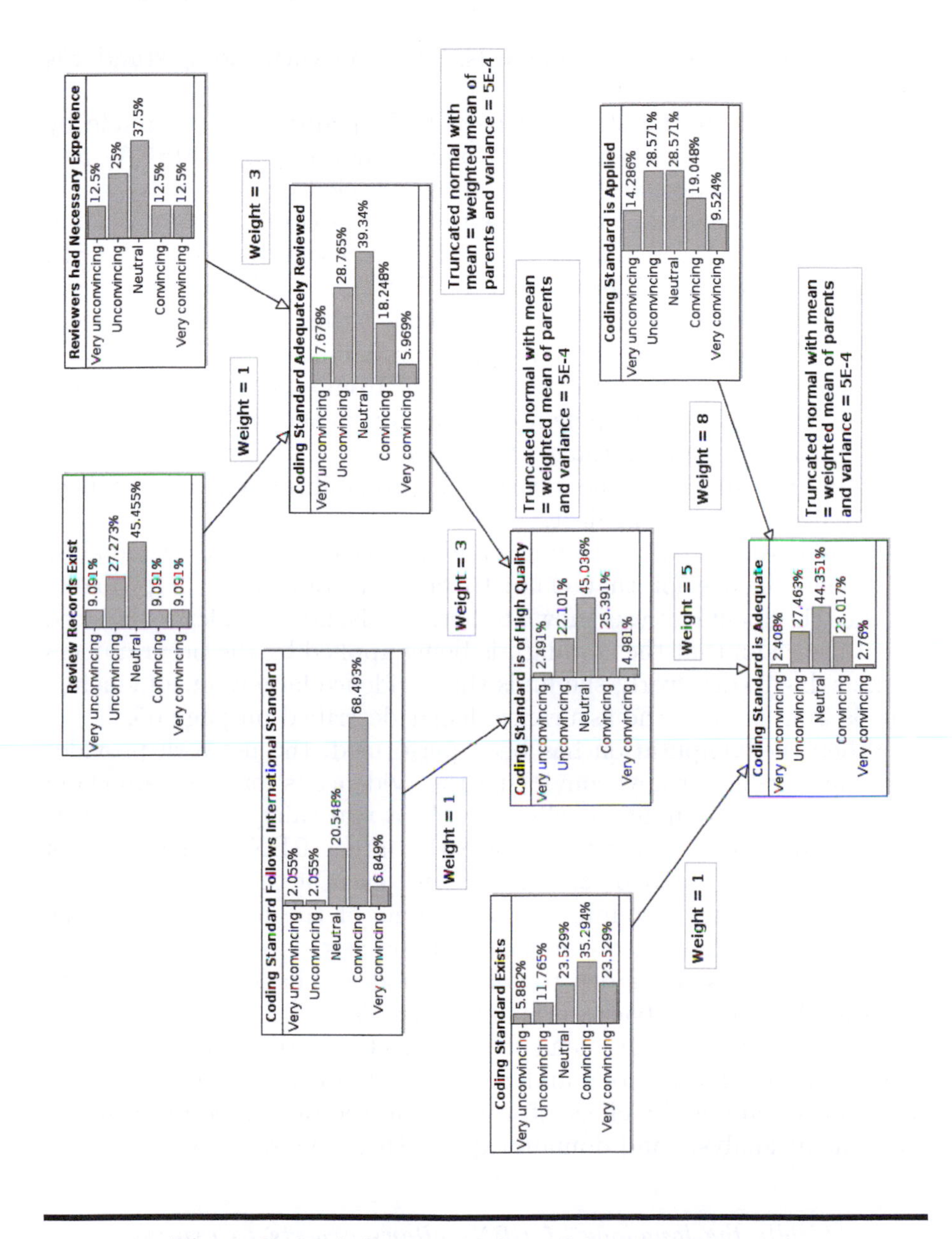

Figure B.4 A fragment of a safety case argument.

Note that the analyst argues that these three conditions are not all equally important: the existence of the coding standard is far less important (has a weight of 1) than whether it is being applied by the programmers (weight of 8). In my experience, there are many companies with published coding standards, but fewer with coding standards that are actually being used.

The claim that the standard is of high quality is to be made by arguing that it follows an international standard (e.g., MISRA) and that it has been adequately reviewed by qualified people.

There are many ways to express the level of confidence in a claim, and in Figure B.4, the analyst has decided to use a discrete scale with five levels. Reference [3] provides a number of examples where experts have found it easier to apply a continuous rather a discrete function. The levels in Figure B.4 are the prior probabilities — the *priors* — the values that the analyst expects to find from experience with other projects within the development company or, in the case of an auditor, from experience across a range of companies.

Figure B.5 illustrates the *a posteriori* probabilities when the analyst has studied the actual project.

It can be seen that the coding standard itself does exist: the analyst finds the evidence for that claim to be very convincing — probably because the document was shown to her. She is, however, less convinced about the claim that the standard is being applied by the programmers in their day-to-day work; she finds that evidence largely unconvincing. This is a measure of the assurance deficit described on page 65.

When the computation has been performed, the network provides a weighted value for how convincing the evidence is for the main claim that the coding standard is effective. In a real safety case argument, this would be fed as one factor into the larger BBN, and a piece of unique evidence would be associated with each of the leaf nodes (i.e., the nodes with no incoming arrows). One of the advantages of the Bayesian network is that evidence and the associated estimate of its credibility can also be added to non leaf nodes. The level of credibility will then be fed both backwards and forwards.

Of course, there is an element of subjective judgement on the analyst's behalf in this work and reference [1] makes some interesting observations about the effect of working on the production of a safety case on the analysts and domain experts themselves:

> *Ideally the language of BBNs allows experts to express their beliefs about a complex problem within a formal probabilistic framework. In practice we expect both that the experts may find the BBN they produced inadequate and that the process itself of building and validating it may change these beliefs. ...[The experts] need to see various (non-obvious) implications of the model to decide whether any of these are counter to their intuitions.*

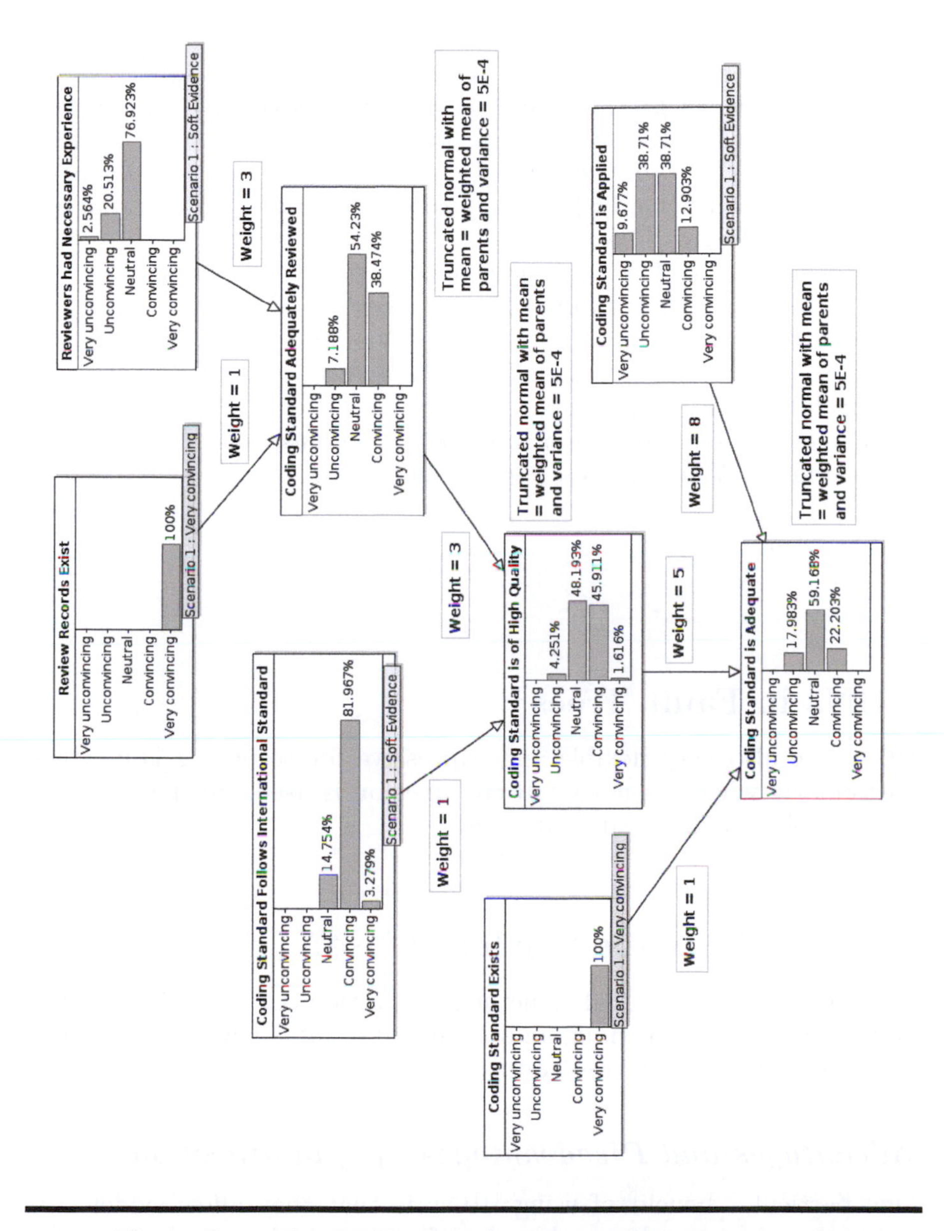

Figure B.5 The BBN after the observations are entered.

These are, of course, positive effects and they imply that the creation of a safety case based on a BBN not only allows domain experts to express their opinions, but also exposes points of inconsistency between the experts and even between conflicting views held by one expert.

Anecdote 26 *A prescriptive standard, such as IEC 61508 (or, to a lesser degree, ISO 26262), lays out what needs to be done to achieve compliance. This is reflected in the checklists, such as that provided by* The CASS Scheme Ltd. *used by some auditors when carrying out an IEC 61508 certification.*

I have suggested on several occasions that it would be better to publish a standard BBN rather than a checklist. A checklist is binary: the item is covered or not, whereas a BBN allows a level of confidence to be recorded. Suitably anonymized, the data from the BBNs could be fed back to improve the priors.

To date, there has been a total lack of enthusiasm on the part of the standards bodies and associated companies.

BBNs in Fault Trees

BBNs are also very flexible and expressive for recording fault trees, particularly when the noisy OR conjunction is used. See page 171 for an example of a Bayesian fault tree.

BBN or GSN for a Safety Case?

Having described the Goal Structuring Notation (GSN) in Appendix A and BBNs in this appendix, an obvious question is which is the better for expressing a safety case argument.

Advantages and Disadvantages of Quantification

One particular benefit of using BBNs is that they allow evidence of differing weights to be used to justify a particular conclusion. For example, using the example from Appendix A, to justify a claim that an algorithm is correct, it may be convenient to present a formal proof of a simplified version of the algorithm, a simulation of the entire algorithm, and some test results. Each of these pieces of evidence supports the claim of correctness of the algorithm, but to different degrees. Similarly, a particular document might have been reviewed independently by two different teams with different skill levels. If both teams report that the document is acceptable, then these two inputs would have different weights.

As illustrated in the examples above, one major difference between a Bayesian and a GSN representation of an argument is that the GSN representation is not quantified. This makes it difficult to balance the worth of different parts of the argument and quantify the "assurance deficit": "Sub arguments A and B are both used to argue that X is true, but we place much more reliance on A rather than on B." A BBN allows us to express *how much* more reliance we place on A.

A lack of quantification also makes it impossible to carry out a sensitivity analysis to determine where it is worth spending more time strengthening evidence to improve the overall confidence in the argument. Figure B.5 clearly indicates that it would be worth spending some time ensuring that the programmers are actually taking notice of the coding standard.

The rebuttal is that using a quantified notation such as BBNs, leads to over confidence in what are, after all, values that are impossible to assign accurately. It is very easy for three statements of "We are about 75% confident" to turn into "The overall confidence is 98.4375%" (perhaps mathematically true but with a completely misleading precision). In particular, it is very difficult to assess good prior probabilities for software systems.

These are strong points in favor of GSN rather than BBN and, in particular, it might be useful to express the argument in GSN format initially and then create a BBN.

The Dirty Washing Argument

Another, to me fallacious, point that is often raised against the BBN is the "dirty washing" argument. A safety case BBN of the type shown in Figure B.5 is ruthless in exposing a company's weaknesses. If an external auditor wants to find a weak area on which to probe, the information is presented here in a very readable form. In the case of Figure B.5, it might be worth exploring inhowfar the programmers are actually using the coding standard. Is this information we want to expose to the auditor?

The first response to that question is provided in the discussion on a company's safety culture on page 4 and the discussion of the target audience for the safety case on page 61. The auditor lies at number five on the list of target audiences on page 61. When used within a strong safety culture, the purpose of the safety case is for the development organization to probe its weaknesses, not to convince an external auditor of a dubious argument. A sophisticated auditor will never expect a product design, implementation and verification to be perfect, and no such argument would be convincing. What the BBN allows us to do is identify the weaknesses and decide whether they need to be improved in order to produce a safe product.

However, I have known an unsophisticated auditor to use this quantification against the analyst: "Why don't you have 100% confidence in this evidence? What's wrong with it?" With such auditors, it might be useful to keep the structure of the BBN, but remove the quantification from the copy presented to the auditor.

References

1. B. Littlewood, L. Strigini, D. Wright, and P.-J. Courtois, "Examination of Bayesian Belief Network for Safety Assessment of Nuclear Computer-based Systems," *DeVa TR No*, vol. 70, 1998.
2. N. Fenton and M. Neil, *Risk Assessment and Decision Analysis with Bayesian Networks.* CRC Press, 2013.
3. M. K. Cowles, *Applied Bayesian Statistics.* New York: Springer, 2013.
4. T. Bayes, "An essay towards solving a problem in the doctrine of chances," *Phil. Trans. of the Royal Soc. of London*, vol. 53, pp. 370–418, 1763.
5. C. Hobbs and M. Lloyd, "The Application of Bayesian Belief Networks to Assurance Case Preparation," in *2012 Safety Critical Systems Symposium*, SSS '12, (Bristol, UK), Safety-Critical Systems Club, 2012.

Appendix C

Notations

I do hate sums. There is no greater mistake than to call arithmetic an exact science. ...If you multiply a number by another number before you have had your tea, and then again after, the product will be different. It is also remarkable that the Post-tea product is more likely to agree with other people's calculations than the Pre-tea result.

Mrs. La Touche of Harristown

Very little mathematics is used in this book, but here and there some mathematical symbols have slipped in. This appendix describes the use of those symbols.

General Symbols

I make use of the symbols listed here without explanation in the text where they appear:

Symbol	**Meaning**
$\forall$	for each
$\exists$	there exists
$\in$	which is a member of
$\mathbb{R}$	the set of real numbers
$\mathbb{R}^n$	the set of all points in n-dimensional space
$a \gg b$	a is much greater than b

As examples, consider the following:

- $\forall x \in \mathbb{R} \quad \exists y \in \mathbb{R} \quad \text{with} \quad y > x$
  This can be read as "for each x, which is a real number, there is a y, which is also a real number, with y greater than x." Put more succinctly, this says that there is no largest real number.
- $(4, 5) \in \mathbb{R}^2$
  The co-ordinates (4, 5) represent a point in two-dimensional space (the plane). Any member of $\mathbb{R}^n$ is known as a *vector*.

Pi and Ip

The use of capital pi to mean multiplication is well known:

$$\prod_{i=3}^{i=7} i = 3 \times 4 \times 5 \times 6 \times 7 = 2520 \tag{C.1}$$

Its associate `ip` is less well known, but is very useful in probability theory:

$$\coprod_{i \in \{0.1, 0.3, 0.7\}} i = 1 - \{(1 - 0.1) \times (1 - 0.3) \times (1 - 0.7)\} \tag{C.2}$$

or, more generally,

$$\coprod_{i=1}^{N} x_i = 1 - \{(1 - x_1) \times (1 - x_2) \times \ldots \times (1 - x_N)\} \tag{C.3}$$

The `ip` notation can be extended into an infix notation* for vectors. Given two vectors $\vec{x} = (x_1, x_2, x_3, \ldots x_N)$ and a similar $\vec{y}$, then

$$\vec{x} \coprod \vec{y} = (1-(1-x_1)(1-y_1), 1-(1-x_2)(1-y_2), \ldots, 1-(1-x_N)(1-y_N)) \tag{C.4}$$

* "Infix" here means that the operator is placed between its operands. Thus, $a + b$ rather than $+(a, b)$.

The Structure Function

A system that consists of N components can be described by a vector of N elements, each representing the state (1 = functioning, 0 = failed) of one component.

Consider the simple system shown at the top of Figure 10.1 on page 133. This has three components, and the system can therefore be represented as a vector $\vec{x} = (x_A, x_B, x_C)$, where x_A, x_B and x_C represent the states of components A, B and C.

The structure function, $\phi(\vec{x})$, for the system is then a function of the x_i:

$$\phi(\vec{x}) = \begin{cases} 0 \text{ if system has failed} \\ 1 \text{ if system is functioning} \end{cases} \quad \text{(C.5)}$$

Thus, for the example in Figure 10.1,

$$\phi(1,1,1) = 1 \quad \text{(C.6)}$$

$$\phi(1,1,0) = 1$$

$$\phi(1,0,0) = 0$$

$$\phi(0,1,1) = 0 \quad \text{(C.7)}$$

Equation C.7, for example, means that when component A is inoperable and components B and C are functioning correctly, the system has failed, whereas Equation C.6 means that when all three components are functioning, then the system is functioning.

Components in Parallel and Series

Given the definition of $\phi(\vec{x})$, it is easy to see that for a set of components in series (as with old-fashioned Christmas tree lights), where all components must operate correctly for the system to operate:

$$\phi(\vec{x}) = x_1 \times x_2 \times \ldots \times x_N = \prod_{i=1}^{N} x_i \quad \text{(C.8)}$$

For components in parallel, where the system continues to operate until all components have failed:

$$\phi(\vec{x}) = 1 - (1 - x_1) \times (1 - x_2) \times \ldots \times (1 - x_N) = \coprod_{i=1}^{N} x_i \tag{C.9}$$

The $\phi(\vec{x})$ notation can be extended to other structures very easily. For a 2oo3 (two out of three) system where any two of the three components must be operating for the system to function:

$$\phi(x_1, x_2, x_3) = x_1 x_2 \coprod x_1 x_3 \coprod x_2 x_3 \tag{C.10}$$

Temporal Logic

The Need for Temporal Logic

There are several problems with the conventional logic made famous by Aristotle.

One was pointed out in the works of Gottlob Frege and Bertrand Russell at the turn of the 20th century: it cannot handle the truth or falsity of statements such as, "The present King of France is bald." At the time of writing, there is no King of France, and so it might be said that this statement is false. But, by Aristotelean logic, if it is false, then its converse, "The present King of France is not bald," must be true. Russell resolved this paradox to general, although not universal, acceptance in reference [1].

Of more importance for this appendix is the failure of conventional logic to handle time-related statements. Thus, the statement "Today is Wednesday," is sometimes true (about 1/7th of the time) and sometimes false.

There are also time-related words that conventional logic doesn't handle: "until," "always," "eventually." Thus what appear to us to be very simple statements:

It *always* will *eventually* be Wednesday.
The system holds the relay open *until* the battery discharges to 20%.

cannot be expressed as logical statements.

Linear Temporal Logic

Temporal logic introduces new operators into conventional logic to handle exactly these types of situation, see Table C.1. *Linear* temporal logic (LTL), in particular, also introduces the concept of "next": χ.

Table C.1 The operators of LTL.

Operator	Meaning	Operator	Meaning
$\neg$	not	$\wedge$	and
$\vee$	or	$\rightarrow$	implies
$\leftrightarrow$	equivalent	$\Box$	always
$\diamond$	eventually	$\bigcup$	until
χ	next (i.e., χA means A is true in the next state)		

This means that it must be possible to represent points in time as positive integers and be able to say that one point in time was before or after another.

With the symbols from Table C.1, we can express both safety and liveness properties of a system (see page 18 for an explanation of these terms):

- A safety property says that the system never does anything bad. This can be expressed as $\Box A$: "It is always true that A". If A is the statement, "a dangerous voltage is detected within 23ms," then $\Box A$ becomes the safety property that a dangerous voltage is *always* detected within 23ms.
- A liveness property says that the system eventually does something good: $\diamond A$. If A is "the elevator stops," then $\diamond A$ is the liveness property that the elevator always eventually stops. This is a weak fairness property.

These symbols may be combined to make more complex statements. For example, $\diamond\Box A$ means that A will eventually become true and the remain true forever, whereas $\Box \diamond A$ means that it is always true that A will eventually be true. This latter statement is true of the statement above regarding Wednesdays: "It is always true that it is eventually Wednesday." Whatever day of the week we are living, it will eventually be Wednesday.

Consider the following sequence of events for the conditions A, B and C:

Condition	Time →										
	0	1	2	3	4	5	6	7	8	9	
A	T	T	T	T	T	F	F	F	F	F	...
B	F	F	F	T	T	F	F	F	F	F	...
C	F	F	T	F	F	T	F	F	T	F	...

It can be seen that, assuming the pattern for C continues as shown, at time 0:

$A \bigcup B$ A is true until B is true

$\Box \diamond C$ C is always eventually true

$$\neg\Box A \quad \text{A is not always true} \tag{C.11}$$

$$\diamond\neg A \quad \text{A is not always true} \tag{C.12}$$

The two expressions in C.11 and C.12 are logically equivalent, both saying "there is a time in the future when A is not true."

As with any language, there are idioms in LTL that it is useful to learn. The Büchi Store as described in reference [2] by Yih-Kuen Tsay *et al,* provides an open repository of useful idioms.

Nonlinear Temporal Logic

LTL is only one of the possible temporal logic formulations. Computational Tree Logic (CTL) is another and allows branching — it is no longer linear. With CTL statements such as:

- On every path (from here) there is some moment when X is true.
- There is at least one path (from here) on which Y is always true.

can be expressed.

There are, however, statements than can be expressed in LTL but not in CTL. One example is concept that a property will eventually become true and then remain true forever, in every possible future; this cannot be expressed in CTL.

One advantage of CTL over LTL is that model checking is generally of polynomial time complexity, whereas LTL is generally of exponential complexity.

Both CTL and LTL are subsets of a larger symbology known as CTL*, and everything that can be expressed in either CTL or LTL can be expressed in CTL*. However, it can be difficult to write in CTL* because of its expressive complexity.

Vector Bases

Linear algebra is only used in one place in this book: when generating basis path test cases on page 251. However, it may be interesting to understand how the basis vectors, for example paths A, B and C in the table on page 253, arise.

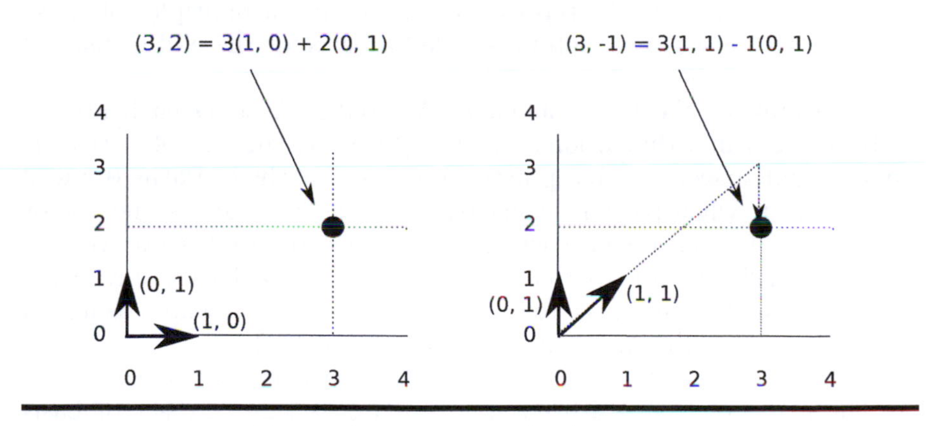

Figure C.1 Different basis vectors.

Consider Figure C.1. On the left-hand side of that figure, I have represented the point (3, 2) using the most commonly used set of basis vectors: (1, 0) and (0, 1). Any point in the two-dimensional plain can be represented as a sum of these two vectors. So, the point (actually vector) that we normally think of as (3, 2) is really

$$(3, 2) = 3 \times (1, 0) + 2 \times (0, 1) \tag{C.13}$$

That is, to reach (3, 2), we move 3 unit vectors along to the right and then 2 unit vectors upwards.

However, (1, 0) and (0, 1), while forming a common basis, are not the only basis. The right-hand side of Figure C.1 shows the same point, now labelled (3, -1). This is because I have chosen to use the two vectors (1,1) and (0, 1) as my basis. In two-dimensional space, two independent vectors will always be needed to define the position of a point precisely. The term "independent" here means that one vector cannot be represented as the sum of multiples of the others. So on the right-hand side of the figure:

$$(3, -1) = 3 \times (1, 1) - 1 \times (0, 1) \tag{C.14}$$

and this can be seen by imagining the (1, 1) vector increased in length 3 times and then a (0, 1) vector being taken away. This illustrates that there is nothing special about (1, 0) and (0, 1). The whole of our two-dimensional geometry could be carried out representing every point as a sum of multiples of (1, 1) and (0, 1).

So, in N-dimensional space, there can be at most N vectors that are independent (i.e., can't be constructed from the others), and every point in the space can be represented by a sum of multiples of those vectors. Such a set of N vectors is called a basis for the N-dimensional space.

Returning to the table on page 253, paths A to D each appear to be vectors in 7-dimensional space. However, this is not a true 7-dimensional space, as some points cannot be reached. Paths 0 and 2 cannot occur when path 1 occurs (see Figure 17.1), and so no vector starting (1, 1, 1 ...) can exist. In fact, when reduced, these vectors are 3-dimensional, and paths A, B and C form a basis: any other reachable point in the space (e.g., path D) can be represented as a sum of multiples of paths A, B and C: see Equation 17.2.

Note that paths A, B and D also form a basis for the space.

The claim of basis path coverage testing is that any set of basis vectors for the control flow graph of a program represents a useful set of test cases.

References

1. B. Russell, "On Denoting," *Mind*, vol. 14, 1905.
2. Y.-K. Tsay, M.-H. Tsai, J.-S. Chang, Y.-W. Chang, and C.-S. Liu, "Büchi Store: an open repository of ω-automata," *International Journal on Software Tools for Technology Transfer*, vol. 15, no. 2, pp. 109–123, 2013.

Index

C, ———

D, ———

J

K

L

T, ———

U, ———

V, ———